Communications
in Computer and Information Science 2167

Series Editors

Gang Li ⓘ, *School of Information Technology, Deakin University, Burwood, VIC, Australia*
Joaquim Filipe ⓘ, *Polytechnic Institute of Setúbal, Setúbal, Portugal*
Ashish Ghosh ⓘ, *Indian Statistical Institute, Kolkata, West Bengal, India*
Zhiwei Xu, *Chinese Academy of Sciences, Beijing, China*

Rationale

The CCIS series is devoted to the publication of proceedings of computer science conferences. Its aim is to efficiently disseminate original research results in informatics in printed and electronic form. While the focus is on publication of peer-reviewed full papers presenting mature work, inclusion of reviewed short papers reporting on work in progress is welcome, too. Besides globally relevant meetings with internationally representative program committees guaranteeing a strict peer-reviewing and paper selection process, conferences run by societies or of high regional or national relevance are also considered for publication.

Topics

The topical scope of CCIS spans the entire spectrum of informatics ranging from foundational topics in the theory of computing to information and communications science and technology and a broad variety of interdisciplinary application fields.

Information for Volume Editors and Authors

Publication in CCIS is free of charge. No royalties are paid, however, we offer registered conference participants temporary free access to the online version of the conference proceedings on SpringerLink (http://link.springer.com) by means of an http referrer from the conference website and/or a number of complimentary printed copies, as specified in the official acceptance email of the event.

CCIS proceedings can be published in time for distribution at conferences or as post-proceedings, and delivered in the form of printed books and/or electronically as USBs and/or e-content licenses for accessing proceedings at SpringerLink. Furthermore, CCIS proceedings are included in the CCIS electronic book series hosted in the SpringerLink digital library at http://link.springer.com/bookseries/7899. Conferences publishing in CCIS are allowed to use Online Conference Service (OCS) for managing the whole proceedings lifecycle (from submission and reviewing to preparing for publication) free of charge.

Publication process

The language of publication is exclusively English. Authors publishing in CCIS have to sign the Springer CCIS copyright transfer form, however, they are free to use their material published in CCIS for substantially changed, more elaborate subsequent publications elsewhere. For the preparation of the camera-ready papers/files, authors have to strictly adhere to the Springer CCIS Authors' Instructions and are strongly encouraged to use the CCIS LaTeX style files or templates.

Abstracting/Indexing

CCIS is abstracted/indexed in DBLP, Google Scholar, EI-Compendex, Mathematical Reviews, SCImago, Scopus. CCIS volumes are also submitted for the inclusion in ISI Proceedings.

How to start

To start the evaluation of your proposal for inclusion in the CCIS series, please send an e-mail to ccis@springer.com.

Mohamed Hamlich · Fadi Dornaika ·
Carlos Ordonez · Ladjel Bellatreche ·
Hicham Moutachaouik
Editors

Smart Applications and Data Analysis

5th International Conference, SADASC 2024
Tangier, Morocco, April 18–20, 2024
Proceedings, Part I

Editors
Mohamed Hamlich
University of Hassan II Casablanca
Casablanca, Morocco

Fadi Dornaika
University of the Basque Country
San Sebastian, Spain

Carlos Ordonez
University of Houston
Houston, TX, USA

Ladjel Bellatreche
ISAE-ENSMA
Poitiers, France

Hicham Moutachaouik
University of Hassan II Casablanca
Casablanca, Morocco

ISSN 1865-0929 ISSN 1865-0937 (electronic)
Communications in Computer and Information Science
ISBN 978-3-031-77039-5 ISBN 978-3-031-77040-1 (eBook)
https://doi.org/10.1007/978-3-031-77040-1

© The Editor(s) (if applicable) and The Author(s), under exclusive license
to Springer Nature Switzerland AG 2024

This work is subject to copyright. All rights are solely and exclusively licensed by the Publisher, whether the whole or part of the material is concerned, specifically the rights of translation, reprinting, reuse of illustrations, recitation, broadcasting, reproduction on microfilms or in any other physical way, and transmission or information storage and retrieval, electronic adaptation, computer software, or by similar or dissimilar methodology now known or hereafter developed.
The use of general descriptive names, registered names, trademarks, service marks, etc. in this publication does not imply, even in the absence of a specific statement, that such names are exempt from the relevant protective laws and regulations and therefore free for general use.
The publisher, the authors and the editors are safe to assume that the advice and information in this book are believed to be true and accurate at the date of publication. Neither the publisher nor the authors or the editors give a warranty, expressed or implied, with respect to the material contained herein or for any errors or omissions that may have been made. The publisher remains neutral with regard to jurisdictional claims in published maps and institutional affiliations.

This Springer imprint is published by the registered company Springer Nature Switzerland AG
The registered company address is: Gewerbestrasse 11, 6330 Cham, Switzerland

If disposing of this product, please recycle the paper.

Preface

The International Conference on Smart Applications and Data Analysis for Smart Cyber-Physical Systems (SADASC 2024) was the conference's fifth edition and it is living up to its reputation as a leading platform for scientists, experts and innovators in the field of cyber-physical systems. SADASC 2024 addressed the different levels of development of data-driven systems and applications, from the source, network and data to the learning and reporting levels of the development lifecycle.

The call for papers resulted in 91 submissions, from which the international program committee carefully selected 30 full papers and 10 short papers, resulting in an acceptance rate of 33%. Each submission was rigorously single blind reviewed by an average of three reviewers, and some were evaluated by as many as four experts. The accepted papers cover a wide range of research areas and deal with both theoretical foundations and practical applications. Notable trends among the accepted submissions include topics such as the Internet of Things, Industry 4.0, smart surveillance, AI-driven systems, networking and sustainable green infrastructures.

At SADASC 2024, we had the honor of hosting two keynote speeches by distinguished experts from Spain and Finland. The opening keynote was given by Sebastian Ventura Soto from the University of Cordoba, Spain, on the topic of "Enhancing maintenance operations through advanced machine learning" followed by Mourad Oussalah from the University of Oulu, Finland, on the topic of "Food Recommender Systems: Bridging the Gap Between Health and User Behavior."

We would like to thank all the authors who contributed to SADASC 2024 and we look forward to their future contributions. We also thank the members of the Program Committee for their thorough and insightful reviews. We thank the user-friendly and customizable OpenReview system for managing the contributions. Finally, we thank the local organizers for their invaluable support.

To the conference participants, we hope that you benefited from the technical sessions, informal meetings and networking opportunities with like-minded people from all over the world. To those who read these proceedings, we hope that the papers will pique your interest and inspire future research projects.

Mohamed Hamlich
Fadi Dornaika
Carlos Ordonez
Ladjel Bellatreche
Hicham Moutachaouik

Organization

General Chairs

Ladjel Bellatreche ISAE-ENSMA, Poitiers, France
Hicham Moutachaouik Hassan II University of Casablanca, Morocco

Program Committee Chairs

Mohamed Hamlich Hassan II University of Casablanca, Morocco
Fadi Dornaika University of the Basque Country UPV/EHU, Spain
Carlos Ordonez University of Houston, USA

Program Committee

Mounir Abid Ibn Zohr University, Morocco
Mohamad Abou Ali University of the Basque Country, Spain, and Lebanese International University, Lebanon
Imad Aboudrar Ibn Zohr University, Morocco
Abdelhafid Aitelmahjoub Hassan II University of Casablanca, Morocco
Jihad Aljaam OUC-LJMU, Qatar
Charbel Aoun NVIDIA, UK
Ignacio Arganda-Carreras University of the Basque Country UPV/EHU, Spain
Mohamed Atibi Hassan II University of Casablanca, Morocco
Abdellah Azmani Abdelmalek Essaadi University, Morocco
Mounir Azmani Abdelmalek Essaadi University, Morocco
Abdelmajid Badri Hassan II University of Casablanca, Morocco
Mostafa Baghouri Hassan II University of Casablanca, Morocco
Abdullah Baradaaji Lebanese International University, Lebanon and University of the Basque Country UPV/EHU, Spain
Fatih Barki Ibn Zohr University, Morocco
Nagore Barrena University of the Basque Country UPV/EHU, Spain
Richard Bearee ENSAM, Lille, France

Djamal Benslimane	Lyon 1 University, France
Jingyun Bi	Beijing University of Posts and Telecommunications, China
Alireza Bosaghzadeh	Shahid Rajaee Teacher Training University, Iran
Jinan Charafeddine	Paris-Saclay Université, France
Adil Chergui	Hassan II University of Casablanca, Morocco
Lahsen-Cherif Iyad	INPT, Morocco
Joseph Constantin	Lebanese University, Lebanon
Octavian Curea	ESTIA, France
Angel Domingo Sappa	Polytechnic University, Ecuador and Computer Vision Center, Spain
Fadi Dornaika	University of the Basque Country UPV/EHU, Spain
Djamel Eddine Boukhari	University of El Oued, Algeria
Moncef Garouani	University of Toulouse, France
Mustapha Hain	Hassan II University of Casablanca, Morocco
Denis Hamad	University of The Littoral Opal Coast, France
Mohamed Hamim	Hassan II University of Casablanca, Morocco
Mohamed Hamlich	Hassan II University of Casablanca, Morocco
Badr Hirchoua	Hassan II University of Casablanca, Morocco
Taha Houda	ISIR, France
Zouilfikar Ibrahim	University of the Basque Country UPV/EHU, Spain
Aissam Jadli	Hassan II University of Casablanca, Morocco
Alvaro Llaria	ESTIA, France
Redouane Majdoul	Hassan II University of Casablanca, Morocco
Mohamed Moutchou	Hassan II University of Casablanca, Morocco
Adel Olabi	ENSAM, Lille, France
Carlos Ordonez	University of Houston, USA
Ilias Ouachtouk	Hassan II University of Casablanca, Morocco
Nabila Rabbah	Hassan II University of Casablanca, Morocco
Mohammed Ramdani	Hassan II University of Casablanca, Morocco
Gilles Regnier	ENSAM, Paris, France
Kaoutar Saidi Alaoui	Ibn Zohr University, Morocco
Sara Sekkate	Hassan II University of Casablanca, Morocco
Ali Siadat	ENSAM, Metz, France
Bouthaina Slika	University of the Basque Country UPV/EHU, Spain
Danyang Sun	École des Ponts ParisTech, France
Mahdi Tavassoli Kejani	Toulouse 3 University, France
Abdelwahed Touati	Hassan II University of Casablanca, Morocco
Viet-Tuan Le	Sejong University, South Korea

Sebastian Ventura	University of Córdoba, Spain
Abdelali Zakrani	Hassan II University of Casablanca, Morocco
Mourad Zegrari	Hassan II University of Casablanca, Morocco

Invited Keynotes

Food Recommender System. Bridging Gap Between Health and User Behavior

Mourad Oussalah

University of Oulu

Abstract. In the digital era, recommender systems played pivotal role in the development of several internet technologies and e-commerce applications where the interaction with the users is central, capitalizing on efficient handling of user's preferences inferred from his past activities or declared profile data. Success of amazon, Netflix and many other service providers are just examples of the efficient and timely eliciting of user's preference to deliver personalized recommendations to users. With the wide spread of food/recipe apps, food industry has also been impacted by this trend. The existence of several food recipe datasets tagged by several communities offer golden opportunities to re-visit the application of standard collaborative/content and hybrid recommender systems to study the impact of several other relevant factors (e.g., food layout, ingredients, nutritional factors, cost, food preparation), and lie down foundation for the development of future recommender systems that better account for health content and user's preference in a way to guide the user towards healthy diet. This keynote talk aims to shed light on the development of food recommender systems, highlighting the key methodological frameworks, current state-of-the-art, challenges and future development. This brings issues of healthy recommender systems, visual recommender systems, emotional-aware recommender system that will be further elucidated. Some ongoing related works at University of Oulu as part of DigiHealth project will be highlighted.

Improving Maintenance Operations with Advance Machine Learning

Sebastián Ventura

Department of Computer Science and Numerical Analysis, University of Córdoba, Córdoba, Spain
sventura@uco.es

Abstract. The industry is experiencing what experts dubbed "The Fifth Industrial Revolution" or "Industry 5.0." This revolution represents the digital transformation of manufacturing production, with a vision that extends beyond mere efficiency and productivity. Instead, Industry 5.0 emphasises the industry's role and contribution to society.

One of the critical advancements in the context of Industry 5.0 is the development of efficient maintenance strategies that utilise artificial intelligence-based models. These strategies involve extracting information from various systems through monitoring, which feeds into artificial intelligence systems. These systems can predict when malfunctions will occur (_predictive systems_) and recommend preventive actions to optimise the operability and availability of the systems (_prescriptive systems_).

Despite the strides made in evolving AI-driven maintenance methodologies, substantial scope exists for enhancement to affirm their reliability. Several challenges persist, encompassing these AI models' sustainability, robustness, resilience, and transparency, among other factors. Sustainability emerges as a paramount concern, underscoring the necessity for models to maintain efficacy over prolonged durations. Robustness is equally imperative, ensuring models adapt to environmental fluctuations and deliver precise forecasts. Resilience is vital for models to recuperate from perturbations swiftly. Moreover, transparency remains crucial to enable user comprehension of the decision-making processes underpinning the models. These challenges will be succinctly examined during this discourse.

The final part of the talk will focus on how the Knowledge Discovery and Intelligent Systems research group at the University of Cordoba addresses some of these challenges. The group is actively working on finding solutions to the issues that arise in implementing AI-based maintenance strategies to advance industrial development further. Specifically, the group is working on improving the interpretability of models through alternative representations, improving data quality through generative models and weak labelling, and improving the sustainability of models through frugal learning and continuous learning techniques.

Contents – Part I

Designing and Modeling

Predicting Lung Infection Severity in Chest X-Ray Images Through
Multi-score Assessment 3
    Bouthaina Slika, Fadi Dornaika, and Karim Hammoudi

Computer Vision and Artificial Intelligent Techniques for Medical Image
Segmentation: An Overview of Technical Aspects and Introduction
to State-of-Art Application 17
    Hanan Sabbar, Hassan Silkan, and Khalid Abbad

Modeling and Managing Product Unavailability Risk in Inventory
Through a Fuzzy Bayesian Network 31
    Ikhlass Boukrouh, Abdellah Azmani, and Samira Khalfaoui

A Binary Particle Swarm Optimization Based Hybrid Feature Selection
Method for Accident Severity Prediction 46
    Mohammed Hamim, Adil Enaanai, Aissam Jadli,
    Hicham Moutachaouik, and Ismail EL Moudden

The Integration of NLP and Topic-Modeling-Based Machine Learning
Approaches for Arabic Mobile App Review Classification 60
    Daniel Voskergian and Faisal Khamayseh

Feature Selection for High-Dimensional Gene Expression Data: A Review ... 74
    Sara Baali, Mohammed Hamim, Hicham Moutachaouik,
    Mustapha Hain, and Ismail EL Moudden

Legal Contract Quality and Validity Assessment Through the Bayesian
Networks 93
    Youssra Amazou, Abdellah Azmani, and Monir Azmani

Optimizing Recommendation Systems in E-Learning: Synergistic
Integration of Lang Chain, GPT Models, and Retrieval Augmented
Generation (RAG) 105
    Qamar EL Maazouzi, Asmaâ Retbi, and Samir Bennani

Data Management

Revolutionizing Skin Cancer Diagnosis: Unleashing AI Precision Through Deep Learning 121
Mohamad Abou Ali, Fadi Dornaika, Ignacio Arganda-Carreras, Hussein Ali, and Malak Karaouni

Fuzzy Bayesian Network Applied to Modeling Vehicles Cooling Systems Failure Risk 139
Soulaimane Idiri, Hafida Khalfaoui, and Abdellah Azmani

Utilization of FMECA to Optimize Predictive Maintenance 153
Maria Eddarhri, Mustapha Hain, and Abdelaziz Marzak

Comprehensive Study on Sentiment Analysis: Insights Into Data Preprocessing, Feature Extraction, and Deep Learning Model Selection 161
Ibtissam Youb and Sebastián Ventura

A Combined AHP-TOPSIS Model for the Selection of Employees for Promotion 175
Loubna Bouhsaien, Abdellah Azmani, and Imane Benallou

Towards Explainable Models: Explaining Black-Box Models 190
Bajja Nisrine, Tabaa Mohamed, and Dufrenois Franck

Advanced NLP and N-Gram Techniques in Financial News Sentiment Analysis: Diverse Machine Learning Approaches 204
Oussama Ndama and El Mokhtar En-Naimi

Comparative Study of Feature Selection Algorithms for Cardiovascular Disease Prediction with Artificial Neural Networks 218
Mohammed Marouane Saim and Hassan Ammor

TinyML and Anomaly Detection

Ensemble Learning for Malware Detection 233
Loubna Moujoud, Meryeme Ayache, and Abdelhamid Belmekki

Enhancing Security Through Data Analysis and Visualization with ELK 246
Zineb Bakraouy, Wissam Abbass, Amine Baina, and Mostafa Bellafkih

Industry 4.0 Efficiency: Predictive Maintenance with TinyML and an Incremental Model 259
Abdelwahed Elmoutaoukkil, Marouane Chriss, Amine Khatib, and Ahmed Mouchtachi

Tiny-ML and IoT Based Early Covid19 Detection Wearable System 268
  Oussama Elallam, Oussama Jami, and Mohamed Zaki

Author Index 281

Contents – Part II

Network Technologies and IOT

ML Based Control in Precision Agriculture: LED Intensity and CO2 Emission Case Study 3
   Pither Gabriel Tene Bermeo, Benaoumeur Senouci, and Jacob Copeland

Industry 4.0: Internet of Things and Cyber-Physical Systems for the Implementation of Pedagogical Factory Digital Twin 16
   Korota Arsène Coulibaly, Alvaro Llaria, Mohamed Hamlich, Octavian Curea, and Rabiae Saidi

Multi-robot Interaction with Mixed Reality for Enhanced Perception 30
   Taha Houda, Ali Amouri, Ayman Beghdadi, and Lotfi Beji

Designing a Firefly Algorithm-Enhanced Protocol for Efficient 3D Heterogeneous Wireless Sensor Networks in Water Quality Monitoring 43
   Ouiam Amenchar, Mostafa Baghouri, Aziz Dkiouak, Saad Chakkor, and Youssef Lagmich

Control, Dynamic Systems and Optimisation

Integrating Novel Circuit Design with Optimization Algorithms for Advanced Parameter Identification in PEM Fuel Cells 55
   Zakaria Kourab, Ismail Ait Taleb, Souad Tayane, Mohamed Ennaji, and Mohamed Haidoury

Human-Robot Collaboration in Remanufacturing: An Application for Computer Disassembly 70
   Soufiane Ameur, Mohamed Tabaa, Mohamed Hamlich, Zineb Hidila, and Richard Bearee

Applied Artificial Neural Networks for Industrial Decision-Making Optimization 85
   Hala Mellouli, Anwar Meddaoui, and Abdelhamid Zaki

Genetic Algorithm-Driven Optimization for Standalone PV/Wind Hybrid Energy Systems Design 93
   Manal Kouihi, Mohamed Moutchou, and Abdelhafid Ait ElMahjoub

Machine Learning Algorithms Based for Demand-Side Energy Forecasting
of an Office Building Loads 107
   Asmae Chakir, Sonia Souabi, and Mohamed Tabaa

Deciphering the Interplay of Precipitation and NDVI for Future Enhanced
Wheat Production Strategies 120
   Sara Bouskour, Mohammed Hicham Zaggaf, and Lhoussain Bahatti

Balancing Assembly Line Based on Lean Management Tools 131
   Youness Hillali, Mourad Zegrari, Najlae Alfathi, and Samir Chafik

A Novel Asymmetrical Multilevel Inverter with Reduced Number
of Switches Regulated by PD-PWM Method 145
   *Sofia Lemssaddak, Mourad Zegrari, Mohamed Tabaa,
and Abdelhafid Ait Elmahjoub*

Exploitation and Exploration

Exploring the Horizon: Challenges and Solutions in Integrating Extended
Reality (XR) into STEM Education 159
   Houda Mouttalib, Mohamed Tabaa, and Mohamed Youssfi

Improving Explainable Matrix Factorization with User-Item Features
for Recommender Systems 173
   Abdelghani Azri, Adil Haddi, and Hakim Allali

Surveying Lightweight Neural Network Architectures for Enhanced
Mobile Performance 187
   Hasnae Briouya, Asmae Briouya, and Ali Choukri

Multi-Agents System in Healthcare: A Systematic Literature Review 200
   Rahma Elkamouchi, Abdelaziz Daaif, and Kamal Elguemmat

Advances in the Security of SDNs: Exploration of Recent ML-Based
Approaches 215
   Ilyasse Mellal, Abderrahime Ait Wakrime, and Khaoula Boukir

Toward an Intelligent Decision Support System for Environmental
Application Using Earth Digital Twin 229
   Feras Al-Obeidat, May AlTaee, and Ali Ben Abbes

Operation and Control of a Five-Level ANPC Inverter 238
   Adnane El-Alami, Radouane Majdoul, and Abdelhafid Ait Elmahjoub

Kinematic Modeling, Optimal Sizing, and Accuracy Analysis of a Compact
Delta Robot 248
  *Said Houmairi, Mohammed Bouaicha, Youssef Elkardaboussi,
  and Mourad Zegrari*

Author Index 265

Designing and Modeling

Predicting Lung Infection Severity in Chest X-Ray Images Through Multi-score Assessment

Bouthaina Slika[1,2], Fadi Dornaika[1,3]([✉]), and Karim Hammoudi[4,5]

[1] University of the Basque Country UPV/EHU, San Sebastian, Spain
bslika001@ikasle.ehu.eus, fadi.dornaika@ehu.eus
[2] Ho Chi Minh City Open University, Ho Chi Minh City, Vietnam
[3] IKERBASQUE, Basque Foundation for Sciene, Bilbao, Spain
[4] Université de Haute-Alsace, IRIMAS, Mulhouse, France
karim.hammoudi@uha.fr
[5] Université de Strasbourg, Strasbourg, France

Abstract. Respiratory illnesses, such as COVID-19, pose a significant health challenge. The prompt and precise identification of pulmonary infections is essential for recognizing patients and deciding on suitable life-saving interventions. In our research, we introduce a multi-task deep learning algorithm crafted to swiftly and effectively evaluate the severity of pulmonary infections in individuals afflicted with COVID-19 or analogous respiratory ailments. Our key contributions entail suggesting a multi-task network utilizing a dual transformer encoder, accompanied by a feature fusion module that feeds two MLP regression heads. Additionally, an online amalgamation of region and score for image augmentation is employed. The consequent model quantifies chest radiographs (CXR) with two scores, characterizing the extent and opacity of infection in the lungs. Evaluation on the RALO dataset demonstrates that our multi-task approach outperforms existing prediction models, exhibiting the minimum mean absolute error and the maximum Pearson correlation coefficient during inference. These results emphasize the effectiveness of inducing two estimation scores in a multi-tasking approach for accurate lung severity diagnosis. The source code is publicly accessible at https://github.com/bouthainas/MViTReg-IP.

Keywords: Computer vision · COVID-19 · Severity quantification · Chest X-ray image analysis · Multi-task Learning · Deep learning

1 Introduction

The emerging coronavirus is a highly infectious virus capable of infecting humans, leading to severe respiratory syndromes [31]. Its rapid global spread has resulted in a pandemic that remains a significant global concern three years later. Contracting SARS-CoV-2 may result in inflammation of the alveoli and accumulation of fluid in the pleura, presenting as challenges in breathing, elevated body temperature, coughing, and symptoms akin to influenza in affected individuals [22]. The impact of COVID-19 extends to the global economic operations [23] and the healthcare system [4]. Molecular diagnostic tests, such as Reverse Transcription Polymerase Chain Reaction (RT-PCR), and radiological imaging tests are crucial for detecting COVID-19 infections [12,16].

While RT-PCR is effective, imaging tests offer a more accessible and rapid screening method, particularly amidst the severe and intricate outbreak of novel coronavirus [35]. In the realm of imaging techniques, chest radiographs are favored over CT scans for diagnosing COVID-19 due to their accessibility, affordability, and lower radiation exposure [26].

Deep learning technology has shown promising results in automated pulmonary diagnosis relying on CXRs, demonstrating excellent performance in detection, classification, localization, monitoring, and severity quantification [28,29]. While numerous studies focus on automatic diagnosis [3,10,20,37], most are classification problems, with limited attention given to lung segmentation [24], infection localization [8], or severity quantification [11,32,33]. Only a few studies have employed multi-task learning to simultaneously perform multiple diagnostic tasks [21,40].

Quantifying infection severity involves assigning scores to lungs based on different factors concerning size, consolidation, density, distribution, and other characteristics indicative of infection in the image. The benefit of multiple scoring methods enhance diagnostic accuracy. However, executing a deep-learning architecture for predicting lung severity faces challenges, including limited access to multi-labeled data, time-consuming and costly data collection, non-uniform distribution of severity levels, and potential concerns about model complexity, overfitting, and interpretability.

Multi-task learning (MTL) addresses the interrelated nature of different quantitative assessments of infections by leveraging potential connections between tasks, resulting in enhanced performance [41]. This approach is also recognized for its efficacy compared to two-stage methodologies of independent predictions. Expanding upon these observations, this study introduces a multi-task learning approach for simultaneous automated quantification of pulmonary infections in chest radiographs. The devised multi-task network incorporates two encoders trained to forecast ratings for quantifying two assessment scores. The geographic extent of infection denotes the dispersion or localization of the illness within the pulmonary tissue, while lung opacity assesses the density of the infected region within the lung. Both metrics offer helpful diagnostic insights. Since CXR image annotations are scalar values, the issue is approached as a regression problem, wherein the model forecasts assessment scores describing the lung infection.

The major contributions of this work are outlined as follows:

– Introduction of a multi-task approach based on the ViTReg-IP encoders capable of concurrently predicting multiple severity quantification scores.

- Application of Combined Lung and Score Replacement utilizing both offline and online augmentation techniques to boost performance.
- Conducting ablation experiments to validate the selected architecture parameters in our investigation.

2 Related Work

While limited single-task models focus on predicting severity scores, this section highlights cutting-edge multi-task approaches. The interest in creating multi-task learning approaches for assessing the seriousness of lung ailments existed even before the emergence of COVID-19. In recent years, the domain of Computer-Aided Diagnosis (CADx) for chest radiographs has seen a notable increase in interest, aiding radiologists in the early detection and diagnosis of diseases. The COVID-19 pandemic has further hastened research and advancement in this field, underscoring the crucial role and the advantages of CADx in the medical field. These approaches encompass CXR image classification and diagnosis, pinpointing abnormalities, assessing the severity of pleural effusion, automating cardiothoracic ratio measurements, lung segmentation, and various applications necessitating the training of Deep Learning models [5,15,17,27,30].

While many research articles have explored AI-driven solutions for monitoring lung diseases and evaluating their severity, few have adopted a multi-task learning approach within a single trainable framework. For instance, Zhou et al. [39] proposed a multi-task learning system aimed at segmenting lung regions and classifying lung nodules as either benign or malignant. Similarly, the authors in [18] devised a deep multi-task learning method capable of detecting lung nodules and segmenting lung regions concurrently from CT images. Moreover, in response to the pandemic, several studies have applied multi-task approaches to research done on COVID-19 analysis. For instance, Amyar et al. [2] introduced a deep multi-task learning architecture to predict COVID-19 lung severity based on CT images, employing a strategy that combines segmentation and classification. In [21], the authors designed a robust algorithm for diagnosing and quantifying the level of COVID-19 relying on the feature corpus extracted from chest radiographs. Additionally, Zhang et al. [40] proposed CXR-Net for actual COVID-19 pneumonia detection, enhancing pixel-level visual understanding utilizing CXR images. Also, Alom et al. [1] introduced a rapid and efficient multi-task deep learning approach for COVID-19 identification. Furthermore, the work in [36] suggested a multi-task semi-supervised learning strategy for the detection of COVID-19.

The ViTReg-IP [33] is a deep learning model derived from the vision transformer developed to assess lung infection severity. In our work [33], we introduced this practical approach for quantifying CXR severity, employing a vision transformer (ViT) alongside a regression head. The ViT decomposes the input CXR image by breaking it into flattened patches and then converting these patches into a latent embedding space. Features derived from this ViT output are directed into a regression head, transforming them into a single score that measures the intensity of the lung infection. This forecasted score value can convey the illness severity. The architecture of the ViTReg-IP deep neural network, as cited in [33], showcased state-of-the-art proficiency in gauging the severity of lung ailments.

Most studies employing multi-task learning were oriented towards the detection, localization, classification, or segmentation of the infection but not quantifying the severity. Additionally, they often applied two distinct tasks conducted jointly, with one task outperforming the other. In contrast, our proposed work demonstrates the training of a model capable of predicting two estimate scores for severity quantification employing the ViTReg-IP model as the backbone structure, both achieving remarkably high performance.

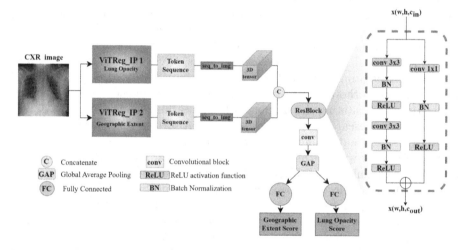

Fig. 1. MViTReg-IP: Two-score prediction model based on multi-task learning.

3 Proposed Methodology

3.1 Multi-task Regressor

In this work, we aim to predict two severity scores concurrently: the geographic extent, which denotes the infected area within the lung, and the lung opacity, which resembles the level of opaqueness associated with the infection as depicted in the CXR image. Our model's primary motivation stems from the observed correlation between both severity scores (geographic extent and lung opacity), implying that jointly estimating these scores could enhance the accuracy of their predictions.

The methodology proposed in this study involves training two separate encoders independently to evaluate the severity of pulmonary ailments. Initially, each encoder is trained to predict one severity score. Subsequently, a novel architecture is trained end-to-end, integrating the two pre-trained encoder models. The resulting architecture, capable of predicting both scores simultaneously, is referred to as the Multi-task Vision Transformer Regression for Infection Prediction (MViTReg-IP), as depicted in Fig. 1.

Two ViTReg-IP-based encoders simultaneously process a CXR image with dimensions $H \times W \times C$. These encoders consist of two pre-trained ViTReg-IP models, each devoid of its regression MLP head. One ViTReg-IP model is pre-trained to predict the geographic extent score, and the other is pre-trained to predict the lung opacity score. Each encoder transforms the input CXR image by dividing it into a sequence of flattened 2D patches, each of size $P \times P$, and the number of patches is $N = (H \times W)/P^2$. These vectorized image patches are then converted into a dimensional embedding space of length D. The last layer of each encoder is a sequence of size $N \times D$. Each layer is reshaped into a 3D tensor of size $(H/P) \times (W/P) \times D$. These features from the two transformers are concatenated into a single tensor. The concatenation is done along the channel dimensions to form a tensor of size $(H/P) \times (W/P) \times 2D$. This tensor undergoes additional processing through a series of layers to improve fusion and extract data more effectively.

The subsequent layers include a ResBlock, comprising two successive convolutional blocks of size 3 by 3 with Batch Normalization (BN) and ReLU activation, followed by a residual connection. The concatenated features delivered by the two transformers are input into this ResBlock, resulting in an output denoted as x_{out}. The next layers consist of a ConvBlock and Global Average Pooling (GAP), producing a 1024-long single vector. Then, this vector fed into two regression heads, each composed of two fully connected layers predicting outcomes simultaneously. Each regression head produces one score for each lung (right and left lungs), with their sum representing the final output score for each branch which is in the range 0–8. One branch predicts the geographic extent, while the other predicts lung opacity. A custom loss function is employed to compare the final predicted values with the ground-truth scores, refining the entire network.

3.2 Data Augmentation

We applied the Combined Lung and Score Replacement augmentation method, involving the exchange of left or right lungs between CXR images of different patients during training. By summing the individual lung scores, we acquired new ground-truth annotations for the resultant images. This strategy was employed both offline and online to expand the training dataset. Offline augmentation increased the dataset size by a factor of three for the RALO dataset, originally comprising 1878 images. This approach enables the model to handle diverse training images, accommodating a broad range of augmented images, including those with noise such as exposure or overlying artifacts. The online augmentation involves replacing lungs in each CXR image within the same batch at each epoch.

3.3 Loss Function

In this study, the employed loss function is a customized function that was derived from the differences between ground truth scores and predicted values for lung opacity and geographic extent. The overall loss for training the proposed architecture is expressed as:

$$\mathcal{L} = \mathcal{L}_{\mathcal{GE}} + \lambda \mathcal{L}_{\mathcal{LO}} \quad (1)$$

where the loss functions for the geographic extent score and the lung opacity score are denoted by $\mathcal{L}_{\mathcal{GE}}$ and $\mathcal{L}_{\mathcal{LO}}$, respectively. The absolute difference between the projected score and the ground truth score is computed for each loss function. The balancing weight λ is assigned to $\mathcal{L}_{\mathcal{LO}}$, and the experimental results indicate optimal performance when λ is set to 0.3.

The CXR images, sized at 224 × 224, underwent training with a batch size of 32, utilizing a learning rate set at 1×10^{-3}, and spanning 60 epochs, with the custom $\mathcal{L}$ as the loss function. The expected durations for training in the conducted tests are outlined in Table 1.

4 Performance Evaluation

4.1 Dataset

The primary aim of this investigation is to assess the viability of Deep Learning-based computer assistance in grading the severity of pulmonary diseases. To achieve this goal, we employed the Radiographic Assessment of Lung Opacity Score (RALO) dataset, by Stony Brook Medicine. This dataset, designed for COVID-19 research, comprises 2373 images, with 1878 images allocated for training and 495 for testing. Two key scoring criteria, geographic extent (GE) and lung opacity (LO), are used for radiological grading. GE determines the degree of infection contribution by ground-glass opacity or consolidation, labelled individually for the right and left lungs from 1 to 4 as the percentage of consolidation increases. The LO score, calculated independently for each lung, ranges from 0 (no opacity) to 4 (total whiteout). The total opacity extent score, derived by summing right and left lung scores, spans from 0 to 8 points [38]. Ground-truth scores, averaged from assessments by two radiologists, fall in the range of 0- 8 with an increment of 0.5.

4.2 Experimental Results and Comparison

To assess the effectiveness of our proposed multi-task approach in evaluating the severity of lung diseases, we conducted a comparative analysis with several existing deep-learning architectures. The performance of our network design demonstrated enhanced sensitivity and improved sensitivity and interpretability in comparison to several established techniques, including COVID-Net [38], COVID-NET-S [37], the feature extraction model by Cohen [7], ResNet50 [13],

Swin Transformer [19], XceptionNet [6], MobileNetV3 [14], and the ViTReg-IP model.

To elevate the severity of pulmonary illness, we applied our model to the refined RALO dataset, compromising 5634 chest radiographs. This dataset provides GE and LO values on a scale of 0 to 8, representing the extent of the illness ranging from mild to critical. In our work, we used both original images and those where the augmentation has been applied. The testing procedures mirror those in our prior study [33], with a distinct online *Combined Lung and Score Replacement* augmentation strategy. This augmentation strategy was consistently applied in all model training sessions. Table 1 offers a concise overview of the performance metrics.

We assess the effectiveness of our proposed model by computing the Mean Absolute Error (MAE) and Pearson Correlation Coefficient (PC) between the predicted values from the model and the ground truth values verified by expert radiologists. An ideal MAE is 0, signifying a precise prediction. The PC ranges from -1 to $+1$, with 0 indicating no correlation and $+1$ representing a perfect linear correlation. Table 1 presents the results of various frameworks in predicting both scores, highlighting the superior performance of our proposed design.

Table 1. Method comparison.

Model	GE MAE ↓	PC ↑	LO MAE ↓	PC ↑	Nb of parameters	Training time*
COVID-NET [38]	4.458	0.549	2.242	0.535	12M	1 hr 20 min
COVID-NET-S [37]	4.698	0.591	2.254	0.529	12M	1 hr 20 min
ResNet50 [13]	1.094	0.688	1.061	0.431	23M	3 hrs
Swin Transformer [19]	0.916	0.817	0.803	0.697	29M	4 hrs
XceptionNet [6]	0.854	0.821	0.768	0.701	23M	3 hrs
Feature Extraction [7]	0.967	0.753	0.865	0.711	2.M	2 hrs
MobileNetV3 [14]	0.847	0.827	0.732	0.738	4.2M	1 hr 20 min
InceptionNet [34]	0.702	0.886	0.609	0.829	24M	3 hr
ViTReg-IP [33]	0.565	0.925	0.510	0.857	5.5M	40 min
MViTReg-IP	**0.531**	**0.938**	**0.462**	**0.881**	11.2M	30 min

*The training time is the total time for the prediction of both scores

4.3 Ablation Studies

The following section presents a comprehensive analysis of the individual components of our multi-task model through a sequence of ablation studies. Firstly, we investigate the impact of several parameters on the performance of both Geographic Extent (GE) and Lung Opacity (LO) scores. Subsequently, we delve into

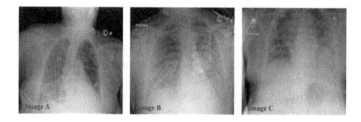

	Predictions					
	Image A		Image B		Image C	
Model	GE	LO	GE	LO	GE	LO
Ground Truth	4	3	7	6	8	4
COVID-NET [38]	2.07	5.02.	5.28	4.23	6.21	6.01
COVID-NET-S [37]	1.38	6.22	5.37	4.53	5.44	6.86
ResNet50 [13]	3.18	3.96	5.23	6.74	4.98	5.57
Feature Extraction [7]	4.76	4.12	8.05	5.11	6.66	5.32
Swin Transformer [19]	3.02	4.08	8.03	5.92	6.69	3.21
XceptionNet [6]	4.95	3.89	7.82	4.42	5.95	5.68
MobileNetV3 [14]	4.87	3.67	7.95	6.98	7.11	3.34
InceptionNet [34]	3.15	3.69	6.28	5.08	6.77	5.01
ViTReg-IP [33]	4.36	3.76	6.54	6.21	7.56	4.41
MViTReg-IP	4.28	3.41	6.85	6.14	7.88	4.32

Fig. 2. Samples of predictions made by the evaluated models on three chest X-ray images

the effect of image pre-processing techniques, including the applied augmentation and lung segmentation. Through these ablation investigations, we aim to gain valuable insights into the mechanisms underlying the effectiveness of our proposed model.

The first study focused on the weighing coefficient λ in the custom loss function $\mathcal{L}$, exploring its impact on model performance for both Geographic Extent (GE) and Lung Opacity (LO) scores (refer to Fig. 3). The results indicated that the optimal performance is achieved when λ is set to 0.3 in the loss function $\mathcal{L} = \mathcal{L}_{\mathcal{GE}} + 0.3\mathcal{L}_{\mathcal{LO}}$.

Another study investigated the influence of augmentation techniques on model performance, specifically employing *Combined Lung and Score Replacement* in both offline and online modes. Table 2 revealed that the model performs best in the case of applying offline and online augmentation simultaneously, leading to the least MAE and the highest PC. Notably, online augmentation particularly benefited the prediction of LO scores. Even in the absence of augmentation, our model demonstrates commendable performance, achieving MAE values of 0.567 for GE and 0.509 for LO, accompanied by PC values of 0.928 for GE and 0.857 for LO. This performance is further underscored by a comparison with the ViTReg-IP single-task model from our previous study [33], which was trained on the same dataset without augmentation and reported higher MAE values of 1.032 for GE, 0.778 for LO, and lower PC values of 0.926 for GE and 0.635 for LO.

In a subsequent ablation study, we examined the impact of varying the size of the training data, testing the whole dataset, half of the dataset, and a quarter of the dataset. The results in Table 3 demonstrated that optimal performance for both our suggested MViTReg-IP model and the InceptionNet (top-performing CNN-based model) is achieved when trained on 100% of the data. Importantly, our suggested method maintained its dominance over the top-performing CNN technique across all three training data subsets, even outperforming the CNN-based method when trained on only 25% of the data size.

We also explored the effect of tuning the encoders in two scenarios: freezing the weights of the encoders versus training the complete architecture end-to-end. Results in Table 4 indicated that training the whole model leads to superior performance.

Furthermore, an ablation study was conducted to determine the batch size percentage represented as α where online *Combined Lung and Score Replacement* is employed during training. Results in Table V revealed that applying this augmentation mode to the entire batch ($\alpha = 100\%$) yields the best results, emphasizing the importance of training the model on sufficient variations of data.

Additionally, we evaluated the impact of lung segmentation on model performance by employing UNet [25] and MA-Net [9]. Results in Table 6 showed that the performance of our MViTReg-IP model remains consistent regardless of whether lung segmentation is performed before training. Notably, for InceptionNet-based models, lung segmentation improved performance, highlighting the effectiveness of self-attention in vision transformers compared to CNN-based models.

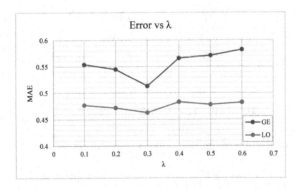

Fig. 3. The model's performance across different values of λ for both scores.

Table 2. Results of ablation studies evaluating the augmentation techniques.

Lung & score replacement		GE		LO	
Offline	Online	MAE ↓	PC ↑	MAE ↓	PC ↑
×	×	0.567	0.928	0.509	0.857
✓	×	0.552	0.931	0.501	0.861
×	✓	0.545	0.933	0.489	0.865
✓	✓	**0.531**	**0.938**	**0.462**	**0.881**

Table 3. Results of ablation studies evaluating the training data size.

	MViTReg-IP				InceptionNet			
	GE		LO		GE		LO	
Training data size	MAE ↓	PC ↑	MAE ↓	PC ↑	MAE ↓	PC ↑	MAE ↓	PC ↑
25%	0.638	0.912	0.566	0.853	0.721	0.859	0.631	0.805
50%	0.599	0.921	0.553	0.863	0.713	0.875	0.617	0.811
100%	**0.531**	**0.938**	**0.462**	**0.881**	0.702	0.886	0.609	0.829

Table 4. Results of ablation studies evaluating the encoder weights.

	GE		LO	
Encoder Weights	MAE ↓	PC ↑	MAE ↓	PC ↑
Frozen	0.603	0.922	0.526	0.867
Trainable	**0.531**	**0.938**	**0.462**	**0.881**

Table 5. Results of ablation studies evaluating the online augmentation technique.

	GE		LO	
α	MAE ↓	PC ↑	MAE ↓	PC ↑
20%	0.618	0.917	0.557	0.859
50%	0.644	0.904	0.566	0.876
70%	0.554	0.928	0.484	0.868
100%	**0.531**	**0.938**	**0.462**	**0.881**

Table 6. Results of applying lung segmentation to input images.

	MViTReg-IP				InceptionNet			
	GE		LO		GE		LO	
Segmentation	MAE ↓	PC ↑	MAE ↓	PC ↑	MAE ↓	PC ↑	MAE ↓	PC ↑
No Segmentation	0.531	0.938	0.426	0.881	0.702	0.886	0.609	0.829
UNet [25]	0.556	0.921	0.455	0.871	0.685	0.852	0.598	0.804
MA-Net [9]	0.586	0.913	0.495	0.869	0.672	0.866	0.577	0.812

5 Discussion

The MViTReg-IP model, proposed in this study, surpasses current single-task deep learning approaches regarding both versatility and performance. This is evident from superior results demonstrated by the MAE and PC scores compared to alternative supervised AI-driven predictive frameworks (refer to Table 1). Notably, our model excels in predicting multiple quantification scores, simultaneously providing assessments for both GE and LO. The minimal training time of our MViTReg-IP translates to reduced computational costs, facilitating swift diagnosis and reduced Time-to-Treatment Initiation (TTI). The experimental results, detailed in Table 1, demonstrate that our MViTReg-IP model achieves the best performance in terms of the MAE metric, attaining a value of 0.531 for GE and 0.462 for LO, the lowest errors among state-of-the-art models. Furthermore, our proposed work's PC metric is the highest, reaching 0.938 for GE and 0.881 for LO scores, indicating a robust positive relationship between predictions and actual severity scores. While ViTReg-IP demonstrates remarkable performance, indicated by its low MAE and high PC, the single-task model's training time for predicting two severity scores exceeds the time required for training the multi-task model performed one time. Figure 2 displays predictions produced by various deep learning models for three CXR images with diverse severity statuses. The table in Fig. 2 illustrates predictions for both GE and LO. The first row shows the ground truth scores of the three images. The MViTReg-IP model aligns the most with radiologist-annotated scores, resulting in the smallest error between actual and predicted scores. Ablation studies, detailed in Subsect. 4.3, highlight the contributions of various parameters, focusing on the loss function, augmentation methods, the size of training data, encoder training status, and lung segmentation. The ablation results are shown in Fig. 3 and Tables 2, 3, 4, 5, and 6, reinforce the adoption of optimal parameters.

6 Conclusion

We introduced a novel multi-task learning model, designed to predict two scores concurrently for assessing the infection severity in lungs, based on CXR images. Through the fine-tuning of pre-trained vision transformers, our method yields precise scores closely aligned with radiologists' annotations. Notably, our approach outperforms state-of-the-art techniques relying on regression models for predicting a single severity score. Experimental outcomes reveal that MViTReg-IP achieves a lower error in predicting the geographic extent score compared to ViTReg-IP alone, and similarly for lung opacity. The integration of our *Combined Lung and Score Replacement* strategy as an online augmentation phase further enhances our model's performance. These resultant scores offer physicians improved objectivity in assessments, constituting a significant achievement of our study.

Acknowledgment. This work is supported in part by grant PID2021-126701OB-I00 funded by MCIN/AEI/10.13039/501100011033 and by "ERDF A way of making Europe".

References

1. Alom, M.Z., Rahman, M., Nasrin, M.S., Taha, T.M., Asari, V.K.: Covid mtnet: Covid-19 detection with multi-task deep learning approaches. arXiv preprint arXiv:2004.03747 (2020)
2. Amyar, A., Modzelewski, R., Li, H., Ruan, S.: Multi-task deep learning based ct imaging analysis for covid-19 pneumonia: Classification and segmentation. Comput. Biol. Med. **126**, 104037 (2020)
3. Apostolopoulos, I.D., Mpesiana, T.A.: Covid-19: automatic detection from x-ray images utilizing transfer learning with convolutional neural networks. Phys. Eng. Sci. Med. **43**, 635–640 (2020)
4. Blumenthal, D., Fowler, E.J., Abrams, M., Collins, S.R.: Covid-19-implications for the health care system (2020)
5. Çallı, E., Sogancioglu, E., van Ginneken, B., van Leeuwen, K.G., Murphy, K.: Deep learning for chest x-ray analysis: a survey. Med. Image Anal. **72**, 102125 (2021)
6. Chollet, F.: Xception: deep learning with depthwise separable convolutions. In: Proceedings of the IEEE Conference on Computer Vision and Pattern Recognition, pp. 1251–1258 (2017)
7. Cohen, J.P., et al.: Predicting covid-19 pneumonia severity on chest x-ray with deep learning. Cureus **12**(7) (2020)
8. Degerli, A., Ahishali, M., Yamac, M., Kiranyaz, S., Chowdhury, M.E., Hameed, K., Hamid, T., Mazhar, R., Gabbouj, M.: Covid-19 infection map generation and detection from chest x-ray images. Health Inf. Sci. Syst. **9**(1), 15 (2021)
9. Fan, T., Wang, G., Li, Y., Wang, H.: Ma-net: a multi-scale attention network for liver and tumor segmentation. IEEE Access **8**, 179656–179665 (2020). https://doi.org/10.1109/ACCESS.2020.3025372
10. Fan, Y., Liu, J., Yao, R., Yuan, X.: Covid-19 detection from x-ray images using multi-kernel-size spatial-channel attention network. Pattern Recogn. **119**, 108055 (2021)
11. Hammoudi, K., Benhabiles, H., Melkemi, M., Dornaika, F., Arganda-Carreras, I., Collard, D., Scherpereel, A.: Deep learning on chest x-ray images to detect and evaluate pneumonia cases at the era of covid-19. J. Med. Syst. **45**(7), 1–10 (2021)
12. Harahwa, T.A., Lai Yau, T.H., Lim-Cooke, M.S., Al-Haddi, S., Zeinah, M., Harky, A.: The optimal diagnostic methods for covid-19. Diagnosis **7**(4), 349–356 (2020)
13. He, K., Zhang, X., Ren, S., Sun, J.: Deep residual learning for image recognition. In: Proceedings of the IEEE Conference on Computer Vision and Pattern Recognition, pp. 770–778 (2016)
14. Howard, A., et al.: Searching for mobilenetv3 (2019). https://doi.org/10.48550/ARXIV.1905.02244, https://arxiv.org/abs/1905.02244
15. Huang, T., Yang, R., Shen, L., Feng, A., Li, L., He, N., Li, S., Huang, L., Lyu, J.: Deep transfer learning to quantify pleural effusion severity in chest x-rays. BMC Med. Imaging **22**(1), 1–11 (2022)
16. Kakodkar, P., Kaka, N., Baig, M.: A comprehensive literature review on the clinical presentation, and management of the pandemic coronavirus disease 2019 (covid-19). Cureus **12**(4) (2020)

17. Li, M.D., et al.: Multi-population generalizability of a deep learning-based chest radiograph severity score for covid-19. Medicine **101**(29) (2022)
18. Liu, W., Liu, X., Li, H., Li, M., Zhao, X., Zhu, Z.: Integrating lung parenchyma segmentation and nodule detection with deep multi-task learning. IEEE J. Biomed. Health Inform. **25**(8), 3073–3081 (2021)
19. Liu, Z., et al.: Swin transformer: hierarchical vision transformer using shifted windows. In: Proceedings of the IEEE/CVF International Conference on Computer Vision, pp. 10012–10022 (2021)
20. Ozturk, T., Talo, M., Yildirim, E.A., Baloglu, U.B., Yildirim, O., Acharya, U.R.: Automated detection of covid-19 cases using deep neural networks with x-ray images. Comput. Biol. Med. **121**, 103792 (2020)
21. Park, S., et al.: Multi-task vision transformer using low-level chest x-ray feature corpus for covid-19 diagnosis and severity quantification. Med. Image Anal. **75**, 102299 (2022). https://doi.org/10.1016/j.media.2021.102299, https://www.sciencedirect.com/science/article/pii/S1361841521003443
22. Pascarella, G., et al.: Covid-19 diagnosis and management: a comprehensive review. J. Intern. Med. **288**(2), 192–206 (2020)
23. Phillipson, J., et al.: The covid-19 pandemic and its implications for rural economies. Sustainability **12**(10), 3973 (2020)
24. Rajaraman, S., Siegelman, J., Alderson, P.O., Folio, L.S., Folio, L.R., Antani, S.K.: Iteratively pruned deep learning ensembles for covid-19 detection in chest x-rays. Ieee Access **8**, 115041–115050 (2020)
25. Ronneberger, O., Fischer, P., Brox, T.: U-Net: convolutional networks for biomedical image segmentation. In: Navab, N., Hornegger, J., Wells, W.M., Frangi, A.F. (eds.) MICCAI 2015. LNCS, vol. 9351, pp. 234–241. Springer, Cham (2015). https://doi.org/10.1007/978-3-319-24574-4_28
26. Rubin, G.D., Ryerson, C.J., Haramati, L.B., Sverzellati, N., Kanne, J.P., Raoof, S., Schluger, N.W., Volpi, A., Yim, J.J., Martin, I.B., et al.: The role of chest imaging in patient management during the covid-19 pandemic: a multinational consensus statement from the fleischner society. Radiology **296**(1), 172–180 (2020)
27. Saiviroonporn, P.: A clinical evaluation study of cardiothoracic ratio measurement using artificial intelligence. BMC Med. Imaging **22**(1), 1–10 (2022)
28. Shi, F., Wang, J., Shi, J., Wu, Z., Wang, Q., Tang, Z., He, K., Shi, Y., Shen, D.: Review of artificial intelligence techniques in imaging data acquisition, segmentation, and diagnosis for covid-19. IEEE Rev. Biomed. Eng. **14**, 4–15 (2020)
29. Shoeibi, A., Khodatars, M., Alizadehsani, R., Ghassemi, N., Jafari, M., Moridian, P., Khadem, A., Sadeghi, D., Hussain, S., Zare, A., et al.: Automated detection and forecasting of covid-19 using deep learning techniques: A review. arXiv preprint arXiv:2007.10785 (2020)
30. Signoroni, A., Savardi, M., Benini, S., Adami, N., Leonardi, R., Gibellini, P., Vaccher, F., Ravanelli, M., Borghesi, A., Maroldi, R., et al.: Bs-net: Learning covid-19 pneumonia severity on a large chest x-ray dataset. Med. Image Anal. **71**, 102046 (2021)
31. Singhal, T.: A review of coronavirus disease-2019 (covid-19). The indian journal of pediatrics **87**(4), 281–286 (2020)
32. Slika, B., Dornaika, F., Hammoudi, K., Hoang, V.T.: Automatic quantification of lung infection severity in chest x-ray images. In: IEEE Statistical Signal Processing (SSP) Workshop. pp. 418–422. IEEE (2023)
33. Slika, B., Dornaika, F., Merdji, H., Hammoudi, K.: Vision transformer-based model for severity quantification of lung pneumonia using chest x-ray images. arXiv preprint arXiv:2303.11935 (2023)

34. Szegedy, C., Liu, W., Jia, Y., Sermanet, P., Reed, S., Anguelov, D., Erhan, D., Vanhoucke, V., Rabinovich, A.: Going deeper with convolutions. In: Proceedings of the IEEE conference on computer vision and pattern recognition. pp. 1–9 (2015)
35. Tahamtan, A., Ardebili, A.: Real-time rt-pcr in covid-19 detection: issues affecting the results. Expert Rev. Mol. Diagn. **20**(5), 453–454 (2020)
36. Ullah, Z., Usman, M., Gwak, J.: Mtss-aae: Multi-task semi-supervised adversarial autoencoding for covid-19 detection based on chest x-ray images. Expert Syst. Appl. **216**, 119475 (2023)
37. Wang, L., Lin, Z.Q., Wong, A.: Covid-net: A tailored deep convolutional neural network design for detection of covid-19 cases from chest x-ray images. Sci. Rep. **10**(1), 1–12 (2020)
38. Wong, A., Lin, Z., Wang, L., Chung, A., Shen, B., Abbasi, A., Hoshmand-Kochi, M., Duong, T.: Towards computer-aided severity assessment via deep neural networks for geographic and opacity extent scoring of sars-cov-2 chest x-rays. Sci. Rep. **11**(1), 1–8 (2021)
39. Wu, B., Zhou, Z., Wang, J., Wang, Y.: Joint learning for pulmonary nodule segmentation, attributes and malignancy prediction. In: 2018 IEEE 15th International Symposium on Biomedical Imaging (ISBI 2018). pp. 1109–1113. IEEE (2018)
40. Zhang, X., Han, L., Sobeih, T., Han, L., Dempsey, N., Lechareas, S., Tridente, A., Chen, H., White, S., Zhang, D.: Cxr-net: A multitask deep learning network for explainable and accurate diagnosis of covid-19 pneumonia from chest x-ray images. IEEE journal of biomedical and health informatics (2022)
41. Zhang, Y., Yang, Q.: A survey on multi-task learning. IEEE Trans. Knowl. Data Eng. **34**(12), 5586–5609 (2021)

Computer Vision and Artificial Intelligent Techniques for Medical Image Segmentation: An Overview of Technical Aspects and Introduction to State-of-Art Application

Hanan Sabbar[1]($\boxtimes$), Hassan Silkan[1], and Khalid Abbad[2]

[1] LAROSERI Laboratory, Computer Science Department, Chouaib Doukkali University, EL Jadida, Morocco
sabbar.h@ucd.ac.ma

[2] Intelligent Systems and Applications Laboratory Science Department, Université Sidi Mohamed Ben Abdellah, Fèz, Morocco
Khalid.abbad@usmba.ac.ma

Abstract. Medical image segmentation is a crucial task in computer vision, playing a fundamental role in applications like diagnosis, treatment planning, and medical research. This article offers a comprehensive survey of various methods employed in medical research for image segmentation. These techniques range from traditional approaches based on thresholds, regions, edges, and clustering, to modern artificial intelligence methods, particularly deep learning techniques. The strengths and limitations of each method are meticulously examined. Furthermore, recent advancements in segmentation methods are scrutinized, emphasizing their potential to enhance both accuracy and efficiency. The study presents results from multiple approaches, accompanied by a detailed analysis of the strengths and weaknesses inherent in the diverse techniques applied to medical image segmentation. This paper focuses on analyzing various architectures used for medical image segmentation, specifically evaluating their performance. It aims to deeply explore the different segmentation methods, offering a comparative perspective on their effectiveness. This study contributes to a better understanding of the applicability of these techniques in the medical field, particularly in computer vision.

Keywords: Segmentation · Medical image · Traditional methods · Computed Tomography (CT) · Machine learning · Deep learning · Computer vision

1 Introduction

The use of image analysis for disease diagnosis has been an established practice for many decades. Today, there are various modalities of medical imaging commonly used in medical practice, such as radiography, MRI, computed tomography (CT), ultrasound, and others. The choice of imaging modality depends on factors such as acquisition speed, image resolution, and patient comfort.

Once a medical image is acquired, a healthcare professional examines it carefully to detect possible diseases and their potential causes. This process can take from a few hours to several days depending on the complexity of the case, involving the participation of expert clinicians and technicians. They assess the size of organs and determine if there are anomalies requiring treatment. All these tasks involve identifying regions of interest, even if segmentation is not always explicitly mentioned. This is why medical image segmentation plays a crucial role in many medical applications, such as the identification and measurement of anomalies, the design of surgical strategies, the monitoring of disease, and significantly facilitating the work of healthcare professionals by identifying areas of interest in medical images. Medical image segmentation can be complex due to inherent challenges such as low contrast, noise, and artifacts in the images. Over the years, various segmentation techniques have been developed to address these problems, and notably, deep learning-based approaches have shown remarkable performance. This article provides a comprehensive review of medical image segmentation techniques, emphasizing their strengths, limitations, and applications in different medical imaging modalities. The choice of specific techniques or algorithms over others depends on the type of image and the nature of the problem. Recent advancements in image segmentation techniques have frequently been the subject of reviews [1,2]. Medical image segmentation techniques can be categorized into two groups: traditional methods based on machine learning and advanced methods based on artificial intelligence. Here is a representation of the predominant medical image segmentation techniques found in each category presented in Fig. 1.

The structure of this article is as follows: Sect. 2 provides a review of existing literature, covering previously conducted related work. Section 3 offers an overview of various traditional architectures used in medical image segmentation. Section 4 delves into the description of the latest architectures based on artificial intelligence for medical image segmentation. Section 5 features a comparative study of different deep learning and traditional architectures. Finally, Sect. 6 concludes the article, discussing future research directions and applications in the field of biomedical image segmentation.

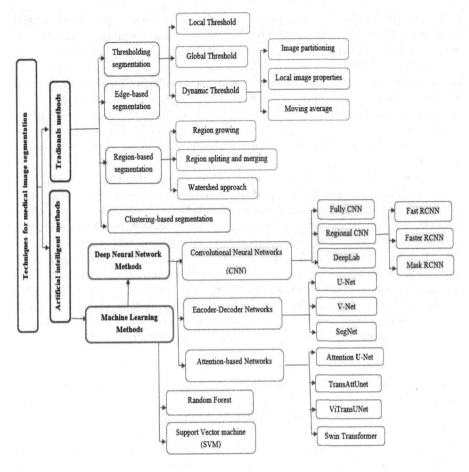

Fig. 1. The techniques for medical image segmentation.

2 Review of Literature

Various methodologies for medical image segmentation have been explored, with Lee. 2007 [3] introducing a statistical approach incorporating morphological operations and Gaussian mixture modeling, demonstrating efficacy in CT image segmentation. Similarly, Ashwani et al. [4] developed a technique based on thresholding and morphology for brain MRI segmentation, validated through CT Angiography. This approach achieved performance ratings of 95.4% for brain MRIs and 95.8% for CT-Angiography, assessed by completeness.

In a recent study, Bhosle et al. 2023 [5] evaluated binary adaptive and Otsu thresholding techniques for lung segmentation in CT images, identifying adaptive thresholding as the superior method with a 78.69% accuracy rate. Binary inverse thresholding followed closely at 75.59%, while Otsu's method, despite its computational simplicity, only achieved 61.70% accuracy due to its lower efficacy in handling images with diverse pixel intensities. This research provides essential insights for selecting the optimal thresholding technique for image segmentation, balancing accuracy, and the particular demands of varying image types.

Zhou et al. 2018 [6] explored the efficacy of an innovative segmentation method for multi-organ detection in CT scans, leveraging a Convolutional Neural Network (CNN) architecture. Their evaluation focused on Mean Accuracy and the Jaccard Similarity Index (JSI), revealing that the method achieved a mean JSI of 79% with a 3D deep CNN and 67% using a 2D deep CNN across seventeen organ types. This indicates the technique's versatility and high performance in segmenting a variety of organs.

Jia et al. 2017 [7], introduced an approach based on Fully Convolutional Networks (FCN) for segmenting histopathology images using deep weak supervision. This method innovatively utilized super-pixels rather than standard pixels, effectively enhancing the preservation of natural tissue boundaries. A key outcome of this approach was its superior performance in segmentation accuracy, as evidenced by an F1 score of 83.6%. This score notably exceeded that of other existing algorithms under weak supervision, marking a significant advancement in the field. Fully convolutional networks (FCN) [8] such as U-Net [9], DeepMedic [10] and holistically nested networks [11,12] have been shown to achieve robust and accurate performance in various tasks including cardiac MR [13], brain tumors [14] and abdominal CT [15,16] image segmentation tasks. Inspired by DenseNet architecture [17].

Ummadi (2022) [18] reviewed U-Net and its derivatives (UNet++, R2UNet, Attention UNet, TransUNet), underscoring their pivotal role in facilitating non-invasive diagnoses through high performance across diverse biomedical segmentation tasks. Inspired by GoogleNet [19,20], Gu et al. [21] introduced CE-Net, which incorporates the inception architecture into medical image segmentation, enhancing feature extraction with atrous convolution for a wider spatial information capture. CE-Net further employs 1×1 convolutions in its feature maps to embody the inception framework, though this complexity adds challenges to model adaptability.

Dosovitskiy et al. [22] introduced the Vision Transformer (ViT), marking a breakthrough in medical image analysis by providing an innovative alternative to conventional convolutional neural networks (CNNs). Originating from advancements in natural language processing, ViT has been effectively applied to medical image segmentation, as evidenced by recent implementations such as TransUnet (2021) [23], Utnet (2021) [24,25], and Swin-unet (2021) [26]. These applications highlight ViT's capability to manage complex interdependencies that exceed CNN's scope. Combining ViT with CNN frameworks is becoming a promising approach for achieving more accurate and efficient segmentation in medical imaging. In the field of medical image segmentation, recent innovations have focused on enhancing accuracy in organ and lesion delineation. Chen et al. (2023) [27] developed TransAttUnet represents a significant advancement in the field of medical image segmentation, an attention-based network that improves semantic segmentation by incorporating multi-level attention and multi-scale connections into the U-Net architecture.

3 Traditional Methods

Traditional medical image segmentation methods encompass a variety of classical image processing and machine learning techniques, each with distinct advantages and limitations. These methods often require manual or semi-automatic intervention, relying on predefined rules, handcrafted features, and mathematical algorithms. Key traditional approaches include thresholding [28], which is simple and practical but can struggle with medical images containing diverse regions, leading to noise and over-segmentation issues. Advanced thresholding techniques like the OTSU method [29] aim to refine this process using local statistical information. Edge-based segmentation [30] accurately detects transitions in image properties but is sensitive to noise, whereas region-based techniques like region growing [31]and the watershed approach [32] group pixels based on similarity, offering diverse segmentation methods but potentially lacking in precision. Clustering-based segmentation groups similar pixels based on intensity or feature similarity. Popular algorithms like K-means or ISODATA [33], fuzzy c-means [34], and the expectation-maximization (EM) algorithm [35]vary in their approach to grouping data, with K-means focusing on mean intensities [36] and fuzzy c-means offering soft segmentations [37]. The EM algorithm assumes Gaussian mixture models to estimate mixture components and posterior probabilities. These traditional methods, while foundational in medical image segmentation, present challenges in handling complex medical image characteristics and often require enhancements for higher accuracy and specificity in diverse medical imaging applications.

4 Intelligence Artificial

With rapid advancements in artificial intelligence, machine learning, and deep learning-based methods have revolutionized the way segmentation is performed. However, with the emergence of deep neural networks, such as Convolutional Neural Networks (CNN) and encoder-decoder Networks, segmentation performance has significantly improved. These deep learning architectures can learn complex features and discriminative patterns from large amounts of data, leading to more accurate and robust segmentation. In the following, we will present an overview of classical machine learning and deep learning methods for medical image segmentation.

4.1 Machine Learning Methods

Machine learning segmentation methods constitute a fundamental aspect of medical image analysis, facilitating the automated extraction and recognition of crucial structures and areas within medical images. The segmentation methods in medical imaging are based on machine learning principles, focusing on Support Vector Machine (SVM) and Random Forest algorithms. SVM, a powerful learning system widely used in pattern recognition, computer vision, and bioinformatics, has demonstrated superior performance compared to traditional classifiers [38]. In medical imaging, SVMs utilize supervised learning to discern complex boundaries between structures, ensuring accurate segmentation of tissues or lesions. Meanwhile, Random Forest, another robust machine learning algorithm for medical imaging, relies on labeled training data, which can be challenging to obtain in medical domains. To address this challenge, semi-supervised learning methods like semi-supervised random forest [39], CoForest [40], and semi-supervised super-pixel method [41] have been introduced, integrating unlabeled data to enhance performance and optimize segmentation accuracy. These techniques represent significant advancements in automating medical image segmentation, enabling precise analysis and diagnosis.

4.2 Deep Neural Network Methods

Currently, deep learning has made significant progress in image segmentation, surpassing conventional methods. The following sections will offer a comprehensive overview of various deep learning techniques for medical image segmentation, encompassing Convolutional Neural Networks (CNNs), encoder-decoder architectures, DeepLab-based segmentation networks, and Transformer models.

Convolutional Neural Networks (CNNs). Convolutional Neural Networks (CNNs) have gained popularity in computer vision and medical image analysis due to their ability to automatically extract relevant features from images, leading to remarkable performance in segmenting anatomical structures and abnormalities in medical images [42]. CNNs (see Fig. 2) consist of three main layers: the convolutional layer, which detects distinct features in images through mathematical operations; the pooling layer, which reduces spatial dimensions without changing depth, reducing computational requirements for subsequent layers; and the fully connected layer, where high-level reasoning and integration of feature responses occur, enabling accurate image analysis. These network architectures have proven effective in medical imaging tasks, revolutionizing the field and contributing significantly to precise image segmentation [42].

As CNN models and architectures have continued to advance, medical image segmentation has achieved unprecedented levels of accuracy and efficiency. Notable deep neural network architectures for image segmentation, including U-Net, V-net, and DeepLab (illustrated in Fig. 1), have played a pivotal role in this progress. These CNN-based segmentation techniques are in a constant state of evolution, continually enhancing segmentation outcomes and broadening the scope of clinical applications. Additionally, recent developments have introduced techniques like TransUNet, TransFuse, MedT, and TransAttUnet, which combine the power of Transformers and CNNs to further elevate the state of medical image segmentation. These hybrid methodologies have demonstrated the potential to address intricate segmentation challenges within the field of medical imaging.

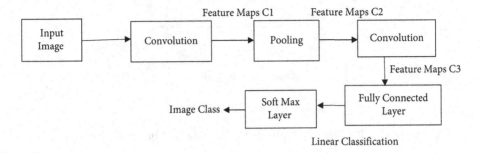

Fig. 2. Convolutional neural network architecture.

U-NET Architecture. Ronneberger et al. [9] introduced the U-Net model at the MICCAI conference in 2015, marking a significant advancement in the application of deep learning for medical imaging segmentation. U-Net, a Fully Convolutional Network (FCN) specifically designed for biomedical image segmentation, comprises an encoder, a bottleneck module, and a decoder. This architecture has gained widespread acceptance for its ability to fulfill the intricate demands of medical image segmentation. Figure 4 illustrates the U-Net structure. Moreover, several foundational U-Net frameworks have been adapted for medical image segmentation, gaining broad implementation. These adaptations of U-Net-related deep learning architectures aim to refine segmentation outcomes by enhancing both the precision and computational efficiency, achieved through modifications in network structures and the integration of novel modules.

Subsequent iterations of U-Net, such as U-Net++, R2U-Net, Attention U-Net, and Trans U-Net, represent progressive enhancements to the original architecture, tailored to improve the accuracy and operational efficiency in medical image segmentation tasks. U-Net++ introduces nested connections to facilitate a more nuanced semantic interpretation and a smoother gradient propagation. R2U-Net merges residual with recurrent connections, enhancing the model's capability in handling temporal sequence data. Attention U-Net incorporates attention mechanisms to concentrate on particular areas of interest, and Trans U-Net amalgamates transformer network elements, boosting performance in complex segmentation tasks. These U-Net variations have shown remarkable efficacy, even when trained on limited datasets, proving their high precision in biomedical segmentation endeavors [9].

Fast R-CNN and Faster R-CNN. A detailed presentation of the strengths of various medical image segmentation techniques based on Fast R-CNN and Faster R-CNN is presented in Table 1.

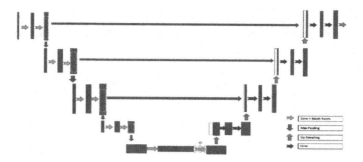

Fig. 3. The structure of U-Net [9].

Table 1. Fast R-CNN and Faster R-CNN in Medical Image Segmentation.

Criteria	Fast R-CNN	Faster R-CNN
Architecture	Utilizes the Region Proposal Network (RPN) and a CNN for feature extraction	Utilizes RPN and a CNN, but with optimizations in the region proposal method
Applications	Used in various medical computer vision applications, including tumor detection and organ segmentation in radiological images	Suited for cases where speed and precision are critical, such as computer-assisted surgery and real-time anomaly detection
Speed	Relatively slow but offers strong performance in accurate segmentation of medical objects.	Improved for increased speed compared to Fast R-CNN, suitable for real-time medical applications
Accuracy	Provides high accuracy in detecting and segmenting complex medical objects.	Provides high accuracy in detecting and segmenting complex medical objects

Transformers. Recent developments in medical image segmentation research have been propelled by innovative neural network architectures. The seminal work by Vaswani et al., which introduced the Transformer by showcasing attention mechanisms, achieved outstanding outcomes across a range of language processing tasks [43]. Chen et al. demonstrated successful segmentation of medical images by integrating Transformers with U-Net, significantly enhancing both localization and contextual understanding [23]. Zhang and colleagues designed TransFuse, a parallel architecture that merges Transformers and CNNs, yielding state-of-the-art performance in diverse medical image segmentation scenarios [44]. The integration of Gated Axial-Attention into MedT by Valanarasu et al. surpassed prior methods in medical image segmentation, setting new benchmarks [25]. To augment semantic segmentation, Chen et al. developed TransAttUnet, employing guided attention to notably advance medical image segmentation efforts [26]. Lin and associates introduced DS-TransUNet by combining Swin Transformer with U-Net, marking a significant evolution in the field [45]. The realm of medical imaging has witnessed a paradigm shift with these advanced neural network architectures coming into play. Among these innovations, ViTransUNet stands out as a groundbreaking development that synergizes the capabilities of Vision Transformers (ViT) with the classic U-Net framework, thereby enhancing the precision and efficiency of image segmentation tasks crucial for various medical applications. Figure 4 depicts the architecture of ViTransUNet.

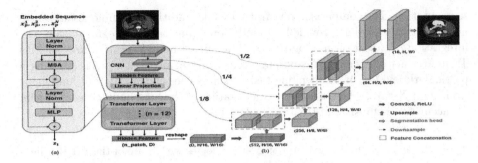

Fig. 4. ViTransUNet architecture.

Incorporating transformers into segmentation frameworks like TransUnet and Swin-Unet has led to notable improvements in segmentation accuracy, especially in demanding tasks such as accurately delineating organs and lesions. This progress is not merely a technological leap; it represents a significant stride towards achieving more precise and minimally invasive diagnostics in healthcare.

5 Comparative Study

In this comparative analysis, we assess the performance of cutting-edge deep learning architectures in contrast to traditional techniques for segmenting lung fields from chest X-rays using the JSRT (Japanese Society of Radiological Technology Database, http://db.jsrt.or.jp/eng.php) and MC (Montgomery) dataset [46]. The findings, presented in Table 2 and represented as percentages, spotlight the remarkable performance of the TransAttUnet model, which achieves an exceptionally high DICE score of 98.88%. This surpasses previous models, both traditional and AI-based, with a notable 2.71% improvement in DICE score over benchmark models like U-Net (96.17%). This improvement underscores the advantages of TransAttUnet's guided encoder-decoder attention and multi-scale skip connections, enabling the model to learn global contextual information and distinctive features that differentiate the lung field from surrounding structures. Furthermore, TransAttUnet consistently outperforms recent approaches such as Attention U-Net (97.59%), FCN (95.1%), and ResUNet++ (97.92%), highlighting its effectiveness in enhancing the quality of detailed segmentation.

The superiority of deep learning over traditional techniques is attributed to its flexibility and ability to adapt to the specifics of medical images. Conventional methods, limited by unchangeable parameters, struggle to handle the complex variability of medical data. In contrast, deep learning adjusts its models for precise segmentation, efficiently leveraging the diversity of features and anomalies present.

This juxtaposition not only validates the advancements brought about by deep learning in medical image analysis but also emphasizes the pivotal role of attention mechanisms in enhancing model sensitivity to relevant features for segmentation. The comparison reveals that while traditional techniques and early neural network models provided a foundational approach for segmentation, the integration of attention mechanisms and advanced neural architectures such as TransAttUnet offers a significant enhancement in segmentation precision. This is particularly evident in challenging tasks such as the segmentation of lung fields from chest X-rays, where the precise delineation of the lung boundaries is crucial for accurate diagnosis and treatment planning.

Table 2. Comparison of Segmentation Methodologies with State-of-the-Art Baselines on the JSRT and MC (Montgomery) Datasets.

Methodology	Dataset	Dice	Accuracy	Recall	Precision
U-Net [9]	JSRT	96.17	98.21	94.94	97.50
FCNN [47]	JSRT & MC	95.1	97.7	95.1	98.0
Encoder-Decoder Structure [48]	JSRT	96.0	–	95.1	–
Improved Segnet [49]	JSRT	–	98.7	–	–
Edge Detection & Morphology [50]	JSRT	–	82.9	–	–
Thresholding [4]	JSRT	–	89.63	88.75	78.76
Fuzzy C-Means (FCM) [4]	JSRT	–	93.34	85.14	92.02
ResUNet [51]	JSRT	97.12	98.64	96.61	97.70
Attention U-Net [52]	JSRT	97.59	98.81	98.82	96.41
UNet++ [53]	JSRT	97.84	98.93	99.28	96.47
ResUNet++ [51]	JSRT	97.92	98.68	98.48	98.48
Swin-Unet [26]	JSRT	97.67	98.71	95.42	98.36
TransUNet [54]	JSRT & MC	–	98.36	–	–
UCTransNet [55]	JSRT	98.32	99.37	–	–
TransAttUnet [56]	JSRT	**98.88**	**98.41**	**98.88**	**99.04**

6 Conclusion and Future Study

In this conclusion and future research outlook, we differentiate between traditional image segmentation techniques, such as thresholding and edge detection, which, despite their simplicity and low training data requirement, struggle with complex images characterized by intensity variations and noise. Conversely, artificial intelligence (AI)-based methods, particularly those employing Convolutional Neural Networks (CNNs) and attention-based networks, exhibit high accuracy. These approaches utilize large datasets to automatically identify relevant features in medical images, proving more adept at handling variations and noise. However, their implementation is complex, requiring significant computational resources and extensive training data, while also being sensitive to data quality and hyperparameter configurations.

Recent advancements, exemplified by models like TransAttUnet, underscore the potential of AI-based approaches in managing complex segmentation tasks and adapting to various medical imaging modalities. The choice between traditional and AI-based methods will depend on the specific needs of the segmentation task, available resources, and the desired performance level.

Future research in medical image segmentation will not only continue to refine data preprocessing techniques, which are vital for enhancing the clarity and precision of image analysis amidst noisy data but will also delve deeper into the architectural evolution of neural networks. A key focus will be the analytical study of the structural advancements from conventional Convolutional Neural Networks (CNNs) to the more recent Transformer models, understanding their unique contributions to the field. This includes examining the incremental improvements brought about by each architecture and dissecting the synergistic effects of their integration. We plan to explore the composite frameworks that couple the locality-sensitive features of CNNs with the dynamic, global contextual awareness of Transformers. The objective is to harness these hybrid models to encapsulate the strengths of both localized and holistic feature extraction methods for superior and more nuanced medical image segmentation. The anticipated outcome is a new generation of segmentation tools that provide not only increased accuracy but also heightened adaptability to diverse medical imaging challenges.

References

1. Zaitoun, N.M., Aqel, M.J.: Survey on image segmentation techniques. Procedia Comput. Sci. **65**, 797–806 (2015)
2. Kumar, S.N., et al.: A voyage on medical image segmentation algorithms. In: Biomedical Research (2017). Special Issue, pp. 1–12
3. Seo, S., Jeong, Y., Stevenson Kenney, J.: A modified CMOS frequency doubler considering delay time matching condition. In: International Symposium on Information Technology Convergence (ISITC 2007), Jeonju, Korea, pp. 392–395. IEEE (2007). https://doi.org/10.1109/ISITC.2007.83
4. Yadav, A.K., et al.: Thresholding and morphological based segmentation techniques for medical images. In: International Conference on Recent Advances and Innovations in Engineering (ICRAIE), pp. 1–5 (2016)
5. Bhosle, S., et al.: Comparative analysis between different lung segmentation techniques. In: Sumathi, A.C., Yuvaraj, N., Ghazali, N.H. (eds.) ITM Web of Conferences, vol. 56, p. 04003 (2023). https://doi.org/10.1051/itmconf/20235604003
6. Zhou, X., et al.: Performance evaluation of 2D and 3D deep learning approaches for automatic segmentation of multiple organs on CT images. In: Medical Imaging 2018: Computer-Aided Diagnosis, vol. 10575, p. 83 (2018)
7. Jia, Z., et al.: Constrained deep weak supervision for histopathology image segmentation. IEEE Trans. Med. Imaging **36**(11), 2376–2388 (2017). https://doi.org/10.1109/TMI.2017.2743464

8. Long, J., Shelhamer, E., Darrell, T.: Fully convolutional networks for semantic segmentation. In: IEEE Conference on Computer Vision and Pattern Recognition (CVPR), pp. 3431–3440 (2015)
9. Ronneberger, O., Fischer, P., Brox, T.: U-Net: convolutional networks for biomedical image segmentation. In: Navab, N., Hornegger, J., Wells, W.M., Frangi, A.F. (eds.) MICCAI 2015. LNCS, vol. 9351, pp. 234–241. Springer, Cham (2015). https://doi.org/10.1007/978-3-319-24574-4_28
10. Kamnitsas, K., et al.: Efficient multi-scale 3D CNN with fully connected CRF for accurate brain lesion segmentation. Med. Image Anal. **36**, 61–78 (2017)
11. Lee, C.Y., et al.: Deeply-supervised nets. In: Artificial Intelligence and Statistics, pp. 562–570 (2015)
12. Xie, S., Tu, Z.: Holistically-nested edge detection. In: Proceedings of the IEEE International Conference on Computer Vision, pp. 1395–1403 (2015)
13. Bai, W., et al.: Human-level CMR image analysis with deep fully convolutional networks. arXiv preprint arXiv:1710.09289 (2017)
14. Kamnitsas, K., et al.: Ensembles of multiple models and architectures for robust brain tumour segmentation. In: Crimi, A., Bakas, S., Kuijf, H., Menze, B., Reyes, M. (eds.) BrainLes 2017. LNCS, vol. 10670, pp. 450–462. Springer, Cham (2018). https://doi.org/10.1007/978-3-319-75238-9_38
15. Roth, H.R., et al.: Spatial aggregation of holistically-nested convolutional neural networks for automated pancreas localization and segmentation. Med. Image Anal. **45**, 94–107 (2018)
16. Roth, H.R., et al.: Hierarchical 3D fully convolutional networks for multi-organ segmentation. arXiv preprint arXiv: 1704.06382 (2017)
17. Li, X., et al.: H-DenseUNet: hybrid densely connected UNet for liver and liver tumor segmentation from CT volumes. arXiv preprint arXiv:1709.07330 (2017)
18. Ummadi, V.: U-Net and Its Variants for Medical Image Segmentation: A Short Review (2022). https://doi.org/10.48550/ARXIV.2204.08470
19. Szegedy, C., et al.: Going deeper with convolutions. In: Proceedings of the IEEE Conference on Computer Vision and Pattern Recognition (CVPR), pp. 1–9, Piscataway, NJ. IEEE (2015)
20. Szegedy, C., et al.: Rethinking the inception architecture for computer vision. In: Proceedings of the IEEE Conference on Computer Vision and Pattern Recognition (CVPR), pp. 2818–2826, Piscataway, NJ. IEEE (2016)
21. Gu, Z., et al.: CE-Net: context encoder network for 2D medical image segmentation. IEEE Trans. Med. Imaging **38**(10), 2281–2292 (2019)
22. Dosovitskiy, A., et al.: An image is worth 16 × 16 words: transformers for image recognition at scale. In: Proceedings of the International Conference on Learning Representations (2019)
23. Chen, J., et al.: TransUNet: transformers make strong encoders for medical image segmentation. arXiv preprint arXiv:2102.04306 (2021)
24. Gao, Y., Zhou, M., Metaxas, D.N.: UTNet: a hybrid transformer architecture for medical image segmentation. In: de Bruijne, M., et al. (eds.) MICCAI 2021. LNCS, vol. 12903, pp. 61–71. Springer, Cham (2021). https://doi.org/10.1007/978-3-030-87199-4_6
25. Valanarasu, J.M.J., et al.: Medical transformer: gated axial-attention for medical image segmentation. arXiv preprint arXiv:2102.10662 (2021)
26. Cao, H., et al.: Swin-Unet: Unet-like pure transformer for medical image segmentation. arXiv preprint arXiv:2105.05537 (2021)

27. Chen, B., et al.: TransAttUnet: multi-level attention-guided U-Net with transformer for medical image segmentation. IEEE Trans. Emerg. Topics Comput. Intell. 8(1), 1–14 (2023). https://doi.org/10.1109/TETCI.2023.3309626
28. Otsu, N.: A threshold selection method from gray-level histograms. IEEE Trans. Syst. Man Cybern. 9(1), 62–66 (1979)
29. Yang, X., Xu, G., Zhou, T.: An effective approach for CT lung segmentation using region growing. J. Phys. Conf. Ser. **2082**, 012001 (2021). https://doi.org/10.1088/1742-6596/2082/1/012001
30. Otsu, N.: Discriminant and least square threshold selection. In: Proceedings of the 4th International Joint Conference on Pattern Recognition, pp. 592–596 (1978)
31. Wu, J., et al.: Texture feature-based automated seeded region growing in abdominal MRI segmentation. In: Proceedings of IEEE International Conference on Biomedical Engineering and Informatics, Sanya, China, vol. 2, pp. 263–267. IEEE (2008)
32. Belaid, L.J., Walid, M.: Image segmentation: a watershed transformation algorithm. Image Anal. Stereol. **28**, 93–102 (2009)
33. Coleman, G.B., Andrews, H.C.: Image segmentation by clustering. Proc. IEEE **5**, 773–85 (1979)
34. Bezdek, J.C., Hall, L.O., Clarke, L.P.: Review of MR image segmentation techniques using pattern recognition. Med. Phys. **20**, 1033–48 (1993)
35. Lei, T., Sewchand, W.: Statistical approach to X-ray CT imaging and its applications in image analysis. II. A new stochastic model-based image segmentation technique for X-ray CT image. IEEE Trans. Med. Imaging **11**(1), 62–69 (1992)
36. Jain, A.K., Dubes, R.C.: Algorithms for Clustering Data, p. 696. Prentice Hall, Englewood Cliffs (1988)
37. Zadeh, L.A.: Fuzzy sets. Inf. Control **8**, 338–353 (1965)
38. Zhou, J., et al.: Nasopharyngeal carcinoma lesion segmentation from MR images by support vector machine. In: 3rd IEEE International Symposium on Biomedical Imaging: Nano to Macro, Arlington, VA, USA, pp. 1364–1367 (2006). https://doi.org/10.1109/ISBI.2006.1625180
39. Leistner, C., et al.: Semi-supervised random forests. In: 2009 IEEE 12th International Conference on Computer Vision, Kyoto, pp. 506–513 (2009). https://doi.org/10.1109/ICCV.2009.5459198
40. Li, M., Zhou, Z.H.: Improve computer-aided diagnosis with machine learning techniques using undiagnosed samples. IEEE Trans. Syst. Man Cybern. **37**, 1088–1098 (2007). https://doi.org/10.1109/TSMCA.2007.904745
41. Gu, L., Zheng, Y., Bise, R., Sato, I., Imanishi, N., Aiso, S.: Semi-supervised learning for biomedical image segmentation via forest oriented super pixels (voxels). In: Descoteaux, M., Maier-Hein, L., Franz, A., Jannin, P., Collins, D.L., Duchesne, S. (eds.) MICCAI 2017. LNCS, vol. 10433, pp. 702–710. Springer, Cham (2017). https://doi.org/10.1007/978-3-319-66182-7_80
42. Milletari, F., Navab, N., Ahmadi, S.-A.: V-Net: fully convolutional neural networks for volumetric medical image segmentation. In: 2016 Fourth International Conference on 3D Vision (3DV), Stanford, CA, USA, pp. 565–571 (2016). https://doi.org/10.1109/3DV.2016.79
43. Vaswani, A., et al.: Attention is all you need. arXiv preprint arXiv:1706.03762 (2023)
44. Zhang, Y., et al.: TransFuse: fusing transformers and CNNs for medical image segmentation. arXiv preprint arXiv:2102.08005 (2021)
45. Lin, A., et al.: DS-TransUNet: dual swin transformer U-Net for medical image segmentation. arXiv preprint arXiv:2106.06716 (2021)

46. Jaeger, S., et al.: Two public chest X-ray datasets for computer-aided screening of pulmonary diseases. Quant. Imaging Med. Surg. **4**(6), 475–477 (2014). https://doi.org/10.3978/j.issn.2223-4292.2014.11.20
47. Rashid, R., Akram, M.U., Hassan, T.: Fully convolutional neural network for lungs segmentation from chest X-rays. In: Campilho, A., Karray, F., ter Haar Romeny, B. (eds.) ICIAR 2018. LNCS, vol. 10882, pp. 71–80. Springer, Cham (2018). https://doi.org/10.1007/978-3-319-93000-8_9
48. Saidy, L., Lee, C.C.: Chest X-ray image segmentation using encoder-decoder convolutional network. In: 2018 IEEE International Conference on Consumer Electronics-Taiwan (ICCE-TW) (2018). https://doi.org/10.1109/ICCE-China.2018.8448537
49. Mittal, A., Hooda, R., Sofat, S.J.W.P.C.: LF-SegNet: a fully convolutional encoder-decoder network for segmenting lung fields from chest radiographs. Wireless Pers. Commun. **101**, 511–529 (2018). https://doi.org/10.1007/s11277-018-5702-9
50. Saad, M.N., et al.: Image segmentation for lung region in chest X-ray images using edge detection and morphology. In: 2014 IEEE International Conference on Control System, Computing and Engineering, Penang, Malaysia (2014)
51. Jha, D.: ResUNet++: an advanced architecture for medical image segmentation. In: IEEE International Symposium on Multimedia (ISM), pp. 225–2255. IEEE (2019)
52. Oktay, O., et al.: Attention UNet: learning where to look for the pancreas. Med. Image Anal. **53**, 2 (2019). https://doi.org/10.1016/j.media.2019.01.012
53. Zhou, Z., Rahman Siddiquee, M.M., Tajbakhsh, N., Liang, J.: UNet++: a nested U-Net architecture for medical image segmentation. In: Stoyanov, D., et al. (eds.) DLMIA/ML-CDS -2018. LNCS, vol. 11045, pp. 3–11. Springer, Cham (2018). https://doi.org/10.1007/978-3-030-00889-5_1
54. Ghali, R., Akhloufi, M.A.: Vision transformers for lung segmentation on CXR images. SN Comput. Sci. **4**(4), 414 (2023). https://doi.org/10.1007/s42979-023-01848-4
55. Li, Y., et al.: Contextual transformer networks for visual recognition. IEEE Trans. Pattern Anal. Mach. Intell. **1**, 3552 (2022). https://doi.org/10.1109/TPAMI.2022.3149543
56. Chen, B., et al.: TransAttUnet: multi-level attention-guided U-Net with transformer for medical image segmentation. arXiv arXiv:2107.05274 (2021)

Modeling and Managing Product Unavailability Risk in Inventory Through a Fuzzy Bayesian Network

Ikhlass Boukrouh[✉], Abdellah Azmani, and Samira Khalfaoui

Intelligent Automation and BioMedGenomics Laboratory, Faculty of Sciences and Techniques of Tangier, Abdelmalek ESSAADI University, Tetouan, Morocco
{ikhlass.boukrouh,samira.khalfaoui}@etu.uae.ac.ma,
a.azmani@uae.ac.ma

Abstract. In the field of inventory management, where important decisions must be made to maintain a balance between product availability and storage costs, adopting innovative approaches for assessing and managing potential risks is essential. This article focuses on analyzing the risk associated with the unavailability of a specific product in stock. It presents a hybrid methodology that combines Bayesian networks for modeling causal relationships between factors influencing stock levels, and fuzzy logic to calculate conditional probabilities of intermediate network nodes, accounting for uncertainty in decision-making in this domain. The developed model's validation is confirmed through the verification of three axioms to ensure its reliability and accuracy. After validation, eight different scenarios were anticipated to assess the factors influencing the reduction of product quantity in stock using a sensitivity analysis. The result of this model was further applied in an event tree analysis to explore the impact of various management strategies on revenue loss.

Keywords: Risk assessment · Inventory management · Fuzzy Bayesian Network · sensitivity analysis · event tree analysis

1 Introduction

Inventory management involves managing and controlling the quantities of products or materials used by an organization in its production and sales operations [1]. The objectives of inventory management include reducing the costs related to storage, handling of inventory [2], and logistics [1], all while meeting customer service requirements, satisfying both customer [2, 3] and supplier needs [3], minimizing theft, increasing profits [1, 3], evaluating the impact of corporate decisions on inventory levels, and achieving a balance between supply and demand [4].

Maintaining a balance between supply and demand aims to ensure the availability of products and avoid inventory issues like understocking and overstocking. Understocking which occurs when demand exceeds supply, potentially causing revenue loss, customer dissatisfaction, and damage to the company's reputation. Conversely, overstocking which often results from overproduction or excessive accumulation, can lead to high storage costs, inefficient use of capital, and product obsolescence. These management errors can cause substantial costs, impacting the financial situation of the company and its reputation in the market.

This research specifically focuses on the challenges associated with unavailability of a specific product in stock and explores strategies to reduce its. The methodology employed combines Bayesian networks (BNs) to model causal relationships between factors affecting unavailability of a product in stock and fuzzy logic (FL) to calculate conditional probabilities (CPs). Sensitivity analysis is applied to test the robustness of the models and ensure the reliability of results, enabling an understanding of the impact of variations and uncertainties on stock management decisions. Furthermore, an event tree is utilized to illustrate and discuss the significance of employing multiple decision-making strategies to reduce the problem of revenue loss due to product unavailability.

This article is structured into four sections, including this introductory one. The second section explores the fuzzy Bayesian networks (FBNs) approach, detailing both its theoretical foundation and practical application. The third section focuses on the event tree approach. Finally, the fourth section presents a conclusion, discusses the limitations of this study, and suggests some future researches.

2 Modeling the Risk of Unavailability of a Product in Stock Using a Fuzzy Bayesian Network Approach

The suggested risk assessment model is designed to estimate the likelihood of unavailability of a product in stock. To accomplish this objective, as illustrated in Fig. 1, the initial step involves conducting a comprehensive investigation to identify the key parameters that may influence the probability of this risk and to establish a causal relationship between them, resulting the creation of a BN. A BN requires CPs, and due to the absence of precise event probabilities, FL has been employed to model uncertain or imprecise human knowledge. By integrating FL theory and expert knowledge, rules were generated and CPs were calculated. This process led to the creation of a FBN model, achieved by combining the causal relationship graph with the CPs calculated by the FL system. Then this approach allowed us to assess product stock availability probabilities across various scenarios and perform sensitivity analysis.

2.1 Bayesian Network Construction

By integrating artificial intelligence and statistics, a BN becomes a graphical probabilistic model that facilitates the representation and utilization of uncertain knowledge [5]. This model represents the dependencies between variables using a directed acyclic graph, while quantifying these dependencies using CPs [6]. Therefore, a BN comprises three key components: 1) nodes representing variables, 2) directed links signifying dependency relationships among these variables, and 3) the conditional probabilities table (CPTs) that assigns specific probabilities to these dependencies. In the node network, the direction of the arrows indicates the relationships: the node at which the arrows point is known as the child, while the node from which the arrows originate is identified as the parent [7].

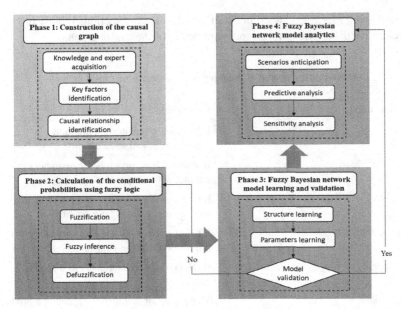

Fig. 1. Methodology of creating a Bayesian network model using fuzzy logic

As shown in Fig. 2, nodes X_1 and X_2 are the parents of node Y_1, and they are the causal factors that affect node Y_1. Conversely, node Y_1 is the child of X_1 and X_2, and its existence is the effect of these two parent nodes. Nodes fall into three categories: root nodes have no parents, intermediate nodes possess both parents and children, and leaf nodes are those without any children. A BN can be designed with a single output, known as Multiple Input Single Output (MISO), or it can have several outputs, which is referred to as Multiple Input Multiple Output (MIMO).

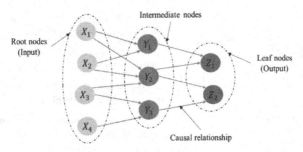

Fig. 2. Structure of a Bayesian network

To assess the risks associated with unavailability of a product in stock, a BN was developed using a comprehensive method and brainstorming sessions to identify key variables and establish dependency relationships between them. After constructing the network, experts in the field conducted an initial validation to verify the reliability and relevance of the model. Additionally, due to the limited historical stock data, CPs were calculated using FL, as explained in detail in the following section. Table 1 provides a comprehensive description of the input parameters of the causal graph.

Table 1. Description of the input parameters of the causal graph.

Parameters name	Description
Packaging	It includes visual design, branding, convenience, and sustainability
Product use	It assesses how easy it is for customers to interact with the product
Price choice	It refers to the process of defining the pricing strategy and options for a product
Product aesthetics	It refers to the visual and sensory aspects of a product's design and appearance
Product quality	It refers to the measurement of a product's durability
Raw materials	The quality of raw materials used in a product
Communication quality	The effectiveness of the exchange of information between suppliers and customers
Responsiveness in updating customer data	It refers to the company's speed and accuracy in updating customer data
Feedbacks analysis frequency	How often feedback is reviewed for insights and improvements
Transportation anomaly scale	It quantifies the degree of anomalies in transportation processes
Storage condition	It refers to the specific environmental under which a product is stored
Purchase volume	It refers to the quantity of products or services that a customer regularly buys
Period of last purchase	It indicates the time passed since a customer's latest purchase from a supplier
Purchase frequency	It is a measure of how often a customer makes purchases from a supplier
Customer presence status	It indicates the customer's availability at the delivery address
Customer address validity	It refers to the accuracy of the customer's provided address
Innovative character	It is the innovation characteristics level of a product proposed by a competitor

(*continued*)

Table 1. (*continued*)

Parameters name	Description
Promotional offer	It is the promotional offer level of a product proposed by a competitor
Market trend	It measures the patterns level in market movement
Pandemics and health crises	It often lead to major market disruptions, increasing the risk of forecasting inaccuracies
Weather situation	
Influential events	It includes consumer-related event and other external events.

Figure 3 illustrates the BN developed to model the factors that influence product unavailability in stock quantity, created using the GeNIe software. In this representation, nodes are color-coded as follows: input nodes appear in yellow, intermediate nodes in light blue, intermediate nodes with a direct impact on the final result are in light purple, and the output node is highlighted in red. Additionally, a unique green node represents a dual role, functioning both as an input and as an intermediate node that directly influence the final result.

2.2 Conditional Probabilities Generation

After the BN is constructed, the next step is calculating the CPTs for each node. This can be achieved in two principal methods: Objective methods, employing learning algorithms based on databases, and subjective methods, involving the utilization of expert knowledge in the field, often gathered through brainstorming sessions. However, no available databases matching all variables in the created BN were available. Consequently, the subjective approach was chosen.

The study faced the challenge of effectively managing expert opinions due to the large number of CPTs that needed to be calculated. To address this, the study implemented a strategy involving FL. This approach begins by converting expert judgments into fuzzy rules and then applies a fuzzy inference system (FIS) to generate the CPTs. This method reduces the frequency of consultations with experts and facilitates the creation of the required probability tables. Several studies have demonstrated the viability of this approach in the context of risk assessment [7–9].

The standard structure for creating CPTs in classical FL models involves multiple interconnected components that manage input variables to produce outputs [10], which include: Fuzzification, fuzzy inference, and defuzzification.

- Fuzzification transforms input variables into a fuzzy subset through the use of fuzzy linguistic variables and membership functions [10, 11]. To implement this, it's essential to define the quadruple $(\mathcal{X}, \mathcal{U}, \mathcal{T}(\mathcal{X}), \mu)$ for each linguistic variable, where: $\mathcal{X}$ represents the name of the variable, $\mathcal{U}$ is the universe of discourse for variable $\mathcal{X}$, $\mathcal{T}(\mathcal{X})$ constitutes the set of linguistic values, which are qualitative descriptors like

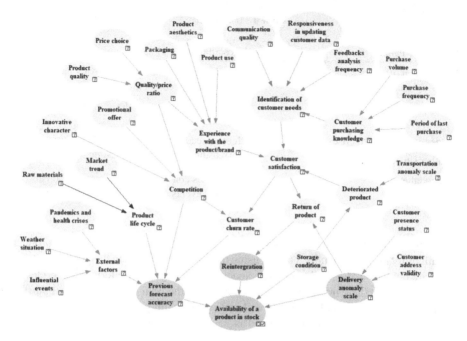

Fig. 3. Bayesian network model for assessing product unavailability in stock risk.

'low', 'medium' or 'high', and μ specifies the membership functions for the various linguistic values of variable $\mathcal{X}$. Various forms of membership functions, such as triangular, trapezoidal, Gaussian, z-shaped, and sigmoid, are utilized in FL [11].

- The fuzzy inference defines the connections between the input and output variables, forming a framework for the fuzzy system's decision-making process. The inference component consists of two main parts: the set of fuzzy rules and the inference engine [12]. The fuzzy rules, formulated based on expert knowledge in natural language, consist of a set of linguistic variables and their associated output representations. These rules are structured as If-Then statements, where the linguistic variables are interconnected using 'and' and 'or' operators. Two popular structures for formulating these fuzzy rules are the Mamdani method and the Takagi-Sugeno-Kang method, also known as the Sugeno method. The primary distinction between these methods lies in how they generate outcomes. In the Mamdani method, the output of each rule is a fuzzy set [13], whereas in the Sugeno method, each rule produces either a linear function based on the input variables or a constant value [14]. The inference engine in the FIS is responsible for formulating final conclusions. It manipulates fuzzy rules and performs implication operations on antecedents to assess their impact on partial conclusions. Subsequently, it aggregates these rules, involving the combination of the partial conclusions from each activated rule (a rule is considered activated if its antecedents have a value that is different from zero) to produce the final result of the inference engine.

- Defuzzification involves converting a fuzzy output into a crisp value. Several methods exist for this, including the centroid method, the bisector method, the mean of maximum, and the largest of maximum [8]. The Sugeno method doesn't require this step since it directly generates a crisp output.

Fuzzification
For this study, excluding the customer's current status node, which was initially represented with binary values 0 (Absent) and 1 (Present), Gaussian membership functions are applied to all remaining nodes, operating within a range of 0 to 1, allowing them to be represented in percentage terms. For example, the "product use" node is defined by the quadruplet $(\mathcal{X}, \mathcal{U}, \mathcal{T}(\mathcal{X}), \mu)$, where:

- $\mathcal{X}$: The name of the variable is "product use".
- $\mathcal{U}$: The universe of this variable is represented by the interval [0, 1].
- $\mathcal{T}(\mathcal{X})$: Linguistic values set, $\{T_1 = easy, T_2 = medium, T_3 = complicated\}$
- μ: Membership functions for each Linguistic value set T_i are defined as follows:

$$\mu_{T_1}(x) = e^{-(x-0.2)^2}, \mu_{T_2}(x) = e^{-(x-0.5)^2}, \mu_{T_3}(x) = e^{-(x-0.8)^2}.$$

Fuzzy Inference
Once all the variables were defined using quadruplets $(\mathcal{X}, \mathcal{U}, \mathcal{T}(\mathcal{X}), \mu)$, the next step involved the creation of a FIS. Sugeno's method was chosen for its efficiency in terms of processing time, which allowed the generation of CPTs without requiring the defuzzification step, thus reducing the potential for information loss.

According to the connections between variables in the RB shown in Fig. 3 and considering the number of their linguistic values, the fuzzy system includes 597 fuzzy rules, resulting in the calculation of 1791 CPs. This research explains the approach for calculating the CPs for the 'customer churn' node, which has two parents: 'customer satisfaction' and 'competition'. This involves 7 rules as presented in Table 2, leading to the calculation of 21 CPs. The same computational method is applied to the other 15 nodes in the system.

Table 2. Fuzzy rules of the "Customer churn rate" node.

Rule	If competition	and customer satisfaction is	Then customer churn rate is
1	Doesn't exist	(bad or medium or good)	Low
2	Exist and bad	Bad	Moderate
3	Exist and bad	Medium	Moderate
4	Exist and bad	Good	Low
5	Exist and good	Bad	High
6	Exist and good	Medium	High
7	Exist and good	Good	Low

The FisPro software was utilized to implement the Sugeno method. The process began by defining the two input nodes and the output node, followed by the development of the rules presented in Table 2. During the creation of the inference engine, Zadeh's operators were employed for the implications of the antecedents and the aggregation of rules: the 'max' operator was used to achieve the union ('or') of different conclusions, while the 'min' operator was used for their intersection ('and') [15]. Considering that "competition" exist and good, and that "customer satisfaction" is bad, the use of the inference engine produced partial conclusions, as illustrated in Fig. 4.

The next step is to aggregate the active rules to determine a unique value for each linguistic variable by using the 'max' operator to achieve the union of the conclusions. Since the probability of a fuzzy set must be different from 0, a value of 0.001 is assigned to states with a null probability.

- Customer churn rate (low) = 0.001
- Customer churn rate (medium) = 0002
- Customer churn rate (high) = max (0.882, 0.044) = 0.882

Since the total of CPs for various states of the "customer churn rate" node must equal 1, the CPs for this node are calculated as follows:

- P (Customer churn rate = low | competition = exist and better and customer satisfaction = bad) = 0.001 / (0.001 + 0.002 + 0.882) = 0.001
- P (Customer churn rate = medium | competition = exist and better and customer satisfaction = bad) = 0.002 / (0.001 + 0.002 + 0.882) = 0.002
- P (Customer churn rate = high | competition = exist and better and customer satisfaction = bad) = 0.001 / (0.001 + 0.002 + 0.882) = 0.997

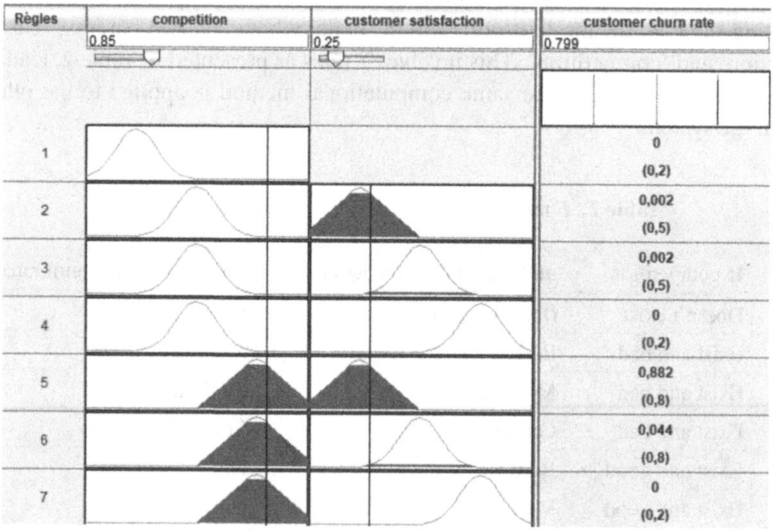

Fig. 4. Fuzzy inference of the node "customer churn rate"

Following these steps, the remaining CPs for the "customer churn rate" node are calculated and presented in Table 3.

Table 3. Conditional probabilities of the node "customer churn rate".

Conditional probabilities					
Rule	Competition	Customer satisfaction	Low	Medium	High
1	Doesn't exist 0.15		0.997	0.002	0.001
2	Exist and bad 0.58	Bad 0.25	0.001	0.890	0.109
3	Exist and bad 0.58	Medium 0.45	0.002	0.889	0.109
4	Exist and bad 0.58	Good 0.75	0.893	0.054	0.054
5	Exist and good 0.85	Bad 0.25	0.001	0.002	0.997
6	Exist and good 0.85	Medium 0.45	0.001	0.997	0.002
7	Exist and good 0.85	Good 0.75	0.993	0.002	0.005

2.3 Model Validation

To ensure the reliability of results obtained from a BN model, validating the model is essential. This study has achieved this through the verification of the three axioms proposed by [16], which have been utilized in the literature by numerous research works, including studies [17–19]. The verification of the axioms was performed across all nodes of the model. As an example of the validation process, consider the case within the child node "Product life cycle".

- **Axiom 1:** Any variation (either an increase or a decrease) in the probability of the parent node must certainly lead to a change in the probability of the child node.

    Axiom 2: A change in the probability distributions of the parent node must result in a consistent effect on the child node.

    - The results of these two axioms for the child node 'Product life cycle' are presented in Table 4, demonstrating their validation. The influences of its parent nodes were assessed by both increasing and decreasing their values.

- **Axiom 3:** The total of the effects of all parent nodes must exceed the effect of each parent separately.

    - The results of applying this axiom to the child node are displayed in Table 5, confirming its validation. The percentage variations indicate that when all parents of the child node are simultaneously increased to 100%, the combined effect is greater that increasing each parent separately.

Table 4. Verification results of axioms 1 and 2.

Axiom 1	Parent node	Child node	Axiom 2	Parent node	Child node
	Raw materials	Product life cycle		Market trend	Product life cycle
20% increase	90%	78,8%	20% increase	92%	60.9%
10% increase	80%	70,3%	10% increase	82%	57.5%
Prior probability	70%	61,6%	Prior probability	72%	48.8%
5% decrease	65%	57%	5% decrease	67%	45.4%
10% decrease	60%	52,9%	10% decrease	62%	42.5%

Table 5. Verification results of axiom 3.

Raw materials	Market trend	Product life cycle	Percentage variations
85%	90%	81.1%	(Initial value)
85%	100%	84.5%	4.17%
100%	90%	95.1%	17.26%
100%	100%	100%	23.30%

2.4 Scenarios Anticipation and Interpretation

After validating the BN model, scenarios were formulated to predict the probability of a product being unavailable in stock. This was achieved by setting the states of root nodes and analyzing their impacts through the propagation of these probabilities. The evaluated scenarios were structured by dividing the input variables into three groups:

1. Supplier service parameters (SSP): Packaging, product use, price choice, product aesthetics, product quality, raw materials, quality of communication, responsiveness in updating customer data, frequency of feedbacks analysis, transportation anomaly scale, and storage conditions.
2. Customer behavior parameters (CBP): Purchase volume, period of last purchase, purchase frequency, customer presence status, and validity of customer address.
3. Operational and environmental parameters (OEP): Innovative character, promotional offer, market trend, influential events, weather situation and pandemics and health crises.

The scenarios were created for each parameter group to represent both favorable (F) and unfavorable (UF) conditions, leading to a total of eight distinct cases, which are presented in Table 6.

Table 6. Predictive scenarios for product stock unavailability.

	Scenario 1	Scenario 2	Scenario 3	Scenario 4	Scenario 5	Scenario 6	Scenario 7	Scenario 8
SSP	F	F	F	F	UF	UF	UF	UF
CBP	F	F	UF	UF	F	F	UF	UF
OEP	F	UF	F	UF	F	UF	F	UF

The probability distribution results for product availability in each scenario are shown in Fig. 5. This illustrates that Scenario 1 has the greatest likelihood of maintaining high product availability, estimated at 93%, with all three variable groups being favorable. In contrast, Scenario 7, which presents the highest probability of significant product unavailability, only the OEP group was favorable.

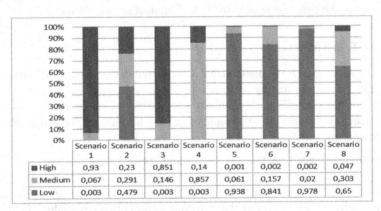

Fig. 5. Probability distribution results for the product availability in stock for each scenario

2.5 Sensitivity Analysis

The purpose of sensitivity analysis is to identify the factors that contribute to the probability of a risk event [20]. This analysis contributes to more effective decision management in order to avoid a risk event.

The predominant parameters in sensitivity analysis for maintaining product consistency in stock are detailed in Fig. 6. These include factors such as poor storage conditions, substandard product quality, inappropriate pricing strategies, and a moderate customer churn rate. These results underscore that supplier service quality significantly influences the probability of product unavailability in stock.

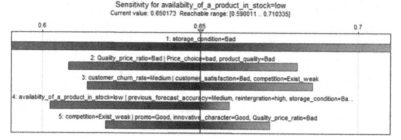

Fig. 6. Sensitivity analysis for product unavailability in stock

3 Predictive Management of Product Unavailability Using Event Tree Analysis

This section presents event tree analysis as a method to prevent revenue loss for suppliers, which is one of the consequences of the risk of product unavailability.

3.1 Event Tree Analysis

Event Tree Analysis (ETA) is a method that uses diagrammatic tree structures to evaluate the consequences of an initiating event in a specific management task. The process starts with evaluating the probability of the initial event, and then branches into various vertical paths representing the success and failure of intermediate events. This highlights their role in either mitigating or aggravating the involved risks [9].

- To construct an event tree, four key steps should be followed:
- Identify the initiating event and its frequency of occurrence.
- Identify mechanisms to prevent the initiating event or to mitigate its consequences, with their effectiveness assessed through probabilities of success/failure. In the event tree, the upper branch usually represents the success of an intermediate event, while the lower branch indicates its failure.
- Identify the consequences.
- Quantify the tree by determining the frequencies of occurrence of different consequences from the identified sequences. The probability of the final result is calculated by multiplying the probabilities associated with each of the branches that lead to that outcome, as follows: $P(R) = \prod_{i=1}^{n} P(B_i)$, where B_i represents the set of branches leading to the result, and n is the number of branches in the set.

In this study, ETA is utilized to assess the risks associated with the unavailability of a product in stock, and its impacts on the supplier's profits. Figure 7 illustrates various scenarios that could result from the initial event of "product unavailability in stock", considering two intermediate events:

- Proactive order management:
  - Success: Rapid restocking of products that are out of stock.
  - Failure: Lag in restocking inventory.

- Alternative product strategies:
  - Success: Effective offering of alternative products.
  - Failure: Lack of alternative products or recommendations.

3.2 Results Interpretation

To forecast the probability of revenue loss for a supplier related to the rate of success or failure of the two key management strategies, Fig. 7 displays the ETA created for this assessment. It indicates that there is a 0.70416 probability of revenue loss when both proactive order management and alternative product strategies fail. Conversely, the event of both strategies succeeding corresponds to a 0.01956 probability of avoiding revenue loss.

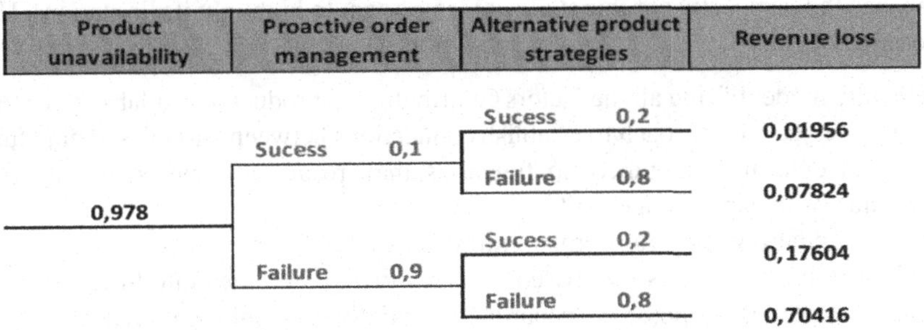

Product unavailability	Proactive order management	Alternative product strategies	Revenue loss
0,978	Success 0,1	Success 0,2	0,01956
		Failure 0,8	0,07824
	Failure 0,9	Success 0,2	0,17604
		Failure 0,8	0,70416

Fig. 7. Event tree showing the probability of revenue loss for a supplier based on the rate of effectiveness of two key management strategies

4 Conclusion, Discussions, Limitations and Future Research

4.1 Conclusion and Discussions

This study evaluated the risk of product unavailability in inventory through a hybrid approach employing a fuzzy Bayesian network. The Bayesian network model was constructed to quantify the influence of root nodes on leaf nodes using a subjective method implemented with GeNIe software. Additionally, fuzzy logic was applied to handle uncertain decision-making by calculating conditional probabilities within the network using the Sugeno method implemented in FisPro software.

To validate the model, three axioms were verified for all nodes in the constructed model. Subsequently, scenario analysis was conducted by categorizing input variables into three groups: supplier service, customer behavior, and operational and environmental parameters. Each group was evaluated under both favorable and unfavorable conditions. The findings indicated that the highest probability of product unavailability in stock occurred when both customer behavior and supplier service were unfavorable, with a probability of 0.978.

To understand the contribution of the model's factors to this risk, sensitivity analysis was performed. This analysis provides valuable insights for industry professionals and decision-makers to develop effective risk prevention strategies. The results demonstrated that the primary influencing factors were a bad "storage condition" a poor "product quality", and inappropriate "price choice".

Product unavailability in stock carries various consequences, one of which is revenue loss. To prevent such revenue loss, a tree event analysis was conducted, considering two intermediate inventories.

The study concluded with a tree event analysis aimed at preventing a consequence of product unavailability in stock, specifically revenue loss, through the utilization of two intermediate management strategies.

4.2 Limitations and Future Research

To enhance the perspective of this study, it is important to highlight its limitations. Here are some of them:

- Difficulty in identifying all the factors contributing to product unavailability in stock.
- Using a subjective approach to establish connections between variables, define fuzzy sets, and determine membership functions may result in errors stemming from inaccuracies in expert judgements.
- Manually entering numerous probabilities can lead to typing errors.
- Anticipation of scenarios was based on categorizing variables into three groups to reduce the number of possible scenarios, potentially overlooking untreated cases.

To overcome these limitations, the future research suggested is:

- Collecting real-world data from industry sources to minimize the utilization of subjective methods.
- Generalizing probability entries to reduce typing errors and save time.

Acknowledgment. The support for this research is provided by the Ministry of Higher Education, Scientific Research, and Innovation, as well as the Digital Development Agency (DDA) and the National Center for Scientific and Technical Research (CNRST) of Morocco, under the Smart DLSP Project - AL KHAWARIZMI IA-PROGRAM.

References

1. Annie Rose Nirmala, D., Kannan, V., Thanalakshmi, M., Joe Patrick Gnanaraj, S., Appadurai, M.: Inventory management and control system using ABC and VED analysis. Mater. Today Proc. **60**, 922–925 (2022). https://doi.org/10.1016/j.matpr.2021.10.315
2. Liu, Y.: A cross-border e-commerce cold chain supply inventory planning method based on risk measurement model. Mob. Inf. Syst. **2022**, 1–9 (2022). https://doi.org/10.1155/2022/6318373
3. Gills, B., Thomas, J.Y., McMurtrey, M.E., Chen, A.N.: The challenging landscape of inventory management. Am. J. Manag. **20**(4), 39–45 (2020)

4. Pourmohammad-Zia, N.: A review of the research developments on inventory management of growing items. J. Supply Chain Manag. Sci. (2021). https://doi.org/10.18757/JSCMS.2021.6122
5. Issa, S.K.: Modélisation d'un outil d'aide à la décision pour la gestion des risques d'incendie dans un édifice (2015)
6. Kaikkonen, L., Parviainen, T., Rahikainen, M., Uusitalo, L., Lehikoinen, A.: Bayesian networks in environmental risk assessment: a review. Integr. Environ. Assess. Manag. **17**(1), 62–78 (2021). https://doi.org/10.1002/ieam.4332
7. Aydin, M., Akyuz, E., Turan, O., Arslan, O.: Validation of risk analysis for ship collision in narrow waters by using fuzzy Bayesian networks approach. Ocean Eng. **231**, 108973 (2021). https://doi.org/10.1016/j.oceaneng.2021.108973
8. Zhang, G.-H., Chen, W., Jiao, Y.-Y., Wang, H., Wang, C.-T.: A failure probability evaluation method for collapse of drill-and-blast tunnels based on multistate fuzzy Bayesian network. Eng. Geol. **276**, 105752 (2020). https://doi.org/10.1016/j.enggeo.2020.105752
9. Khalfaoui, H., Azmani, A., Farchane, A., Safi, S.: Symbiotic combination of a Bayesian network and fuzzy logic to quantify the QoS in a VANET: application in logistic 4.0. Computers **12**(2), 40 (2023). https://doi.org/10.3390/computers12020040
10. Zhang, Y., et al.: Takagi-Sugeno-Kang fuzzy system fusion: a survey at hierarchical, wide and stacked levels. Inf. Fusion **101**, 101977 (2024). https://doi.org/10.1016/j.inffus.2023.101977
11. Samavat, T., et al.: A comparative analysis of the Mamdani and Sugeno fuzzy inference systems for MPPT of an Islanded PV system. Int. J. Energy Res. **2023**, 1–14 (2023). https://doi.org/10.1155/2023/7676113
12. Théorêt, C.: Élaboration d'un logiciel d'enseignement et d'application de la logique floue dans un contexte d'automate programmable. École de technologie supérieure (2009)
13. Mamdani, E.H.: Advances in the linguistic synthesis of fuzzy controllers. Int. J. Man-Mach. Stud. **8**(6), 669–678 (1976). https://doi.org/10.1016/S0020-7373(76)80028-4
14. Sugeno, M., Kang, G.T.: Structure identification of fuzzy model. Fuzzy Sets Syst. **28**(1), 15–33 (1988). https://doi.org/10.1016/0165-0114(88)90113-3
15. Zadeh, L.A.: Fuzzy sets. Inf. Control **8**(3), 338–353 (1965). https://doi.org/10.1016/S0019-9958(65)90241-X
16. Jones, B., Jenkinson, I., Yang, Z., Wang, J.: The use of Bayesian network modelling for maintenance planning in a manufacturing industry. Reliab. Eng. Syst. Saf. **95**(3), 267–277 (2010). https://doi.org/10.1016/j.ress.2009.10.007
17. Chen, Y., Tian, Z., He, R., Wang, Y., Xie, S.: Discovery of potential risks for the gas transmission station using monitoring data and the OOBN method. Reliab. Eng. Syst. Saf. **232**, 109084 (2023). https://doi.org/10.1016/j.ress.2022.109084
18. Göksu, B., Yüksel, O., Şakar, C.: Risk assessment of the ship steering gear failures using fuzzy-Bayesian networks. Ocean Eng. **274**, 114064 (2023). https://doi.org/10.1016/j.oceaneng.2023.114064
19. Guo, C., Wu, W.: Fuzzy dynamic Bayesian network based on a discrete aggregation method for risk assessment of marine nuclear power platform hinge joints accidents. Appl. Ocean Res. **138**, 103674 (2023). https://doi.org/10.1016/j.apor.2023.103674
20. Lin, S.-S., Zhou, A., Shen, S.-L.: Multi-status Bayesian network for analyzing collapse risk of excavation construction. Autom. Constr.. Constr. **158**, 105193 (2024). https://doi.org/10.1016/j.autcon.2023.105193

A Binary Particle Swarm Optimization Based Hybrid Feature Selection Method for Accident Severity Prediction

Mohammed Hamim[1](✉) [iD], Adil Enaanai[2], Aissam Jadli[1], Hicham Moutachaouik[1], and Ismail EL Moudden[3] [iD]

[1] AICSE Laboratory, University Hassan II, Casablanca, Morocco
mohamed.hamim@gmail.com
[2] Department of Computer Science, Faculty of Science, Tetouan, Morocco
[3] Eastern Virginia Medical School, EVMS-Sentara Healthcare Analytics and Delivery Science, Norfolk, USA

Abstract. According to recent reports, road traffic injuries are the leading cause of death among children and young adults. Various systems and strategies have been designed to reduce accident severity. With the development of data mining tools, the use of big traffic data and machine learning techniques holds potential for implementing effective road safety strategies. Using a dataset collected from Addis Ababa, Ethiopia, our research introduces an innovative severity prediction system integrating a hybrid Feature Selection (FS) approach, named OWABPSO, with machine learning algorithms. The OWABPSO approach combines a One-Way-ANOVA-based filter method with a Binary Particle Swarm Optimization-based wrapper method. Six algorithms, including K-Nearest Neighbors, Random Forest, Decision Tree, Light Gradient Boosting Machine, Artificial Neural Network, and Extreme Gradient Boosting, are proposed for severity prediction. Experimental outcomes of this work demonstrate that, compared to state-of-the-art methods, by combining our FS approach with Decision Tree-based classifiers, we achieved competitive results. Our study presents an effective integration of FS approaches in predicting accident severity levels, thus contributing to advanced road safety strategies.

Keywords: Road Traffic Accidents · Feature Selection · Binary Particle Swarm Optimization · Machine Learning

1 Introduction

Road traffic accidents (RTAs) remain a significant factor in personal injury, property loss, and have far-reaching effects on the economy and healthcare system. An RTA can be defined as an incident involving at least one road vehicle on an open public road, resulting in injuries or fatalities for at least one person. A recent WHO (World Health Organization) publication (2022) reports approximately 1.3 million annual premature deaths due to road traffic accidents, with an additional 20 to 50 million people suffering non-fatal injuries, often resulting in permanent disabilities [1]. These injuries also impose a significant economic loss on individuals, families, and nations, including expenses related to medical care and the impact on productivity. This results in costs that reach up to 3% of the Gross Domestic Product (GPD) for most countries [1]. Considering these facts, the implementation of an effective strategy for preventing RTAs has become an urgent imperative for most countries. With the development of data mining tools, accidents can be identified and investigated using various approaches, thereby aiding authorities in developing strategies to reduce the high number of road injuries and fatalities, and thus, to minimize their costs in terms of GPD.

This study aims to develop a system for predicting the severity of RTAs. In this regard, the primary features of our proposed system are outlined as follows:

- Employing an innovative Machine Learning (ML)-based Imputation technique.
- Extracting the most informative features by employing an innovative OWABPSO-based FS technique. The OWABPSO approach comprises a two-step process, combining the One-Way-ANOVA-based filter method with a Binary Particle Swarm Optimization-based wrapper method.
- Implementing the classification process by suggesting six ML algorithms: RF (Random Forest), KNN (k-Nearest Neighbors), ANN (Artificial Neural Network), DT (Decision Tree), LGBM (Light Gradient Boosting Machine), and XGB (Extreme Gradient Boosting) algorithms.
- Assessing the performance of the classification model by employing different evaluation metrics such as Accuracy, F1-score, Precision, Recall, and AUC (Area Under the ROC - Receiver Operating Characteristic - Curve).

This paper is structured as follows. The next section explores relevant studies conducted in the context of predicting the severity of road traffic accidents. Section 3 discusses the methodology employed in the present study. The results of the experiments and the analysis of the effectiveness of our prediction system are discussed in Sect. 4. The final section presents a conclusion and some perspectives.

2 State of the Arts

Due to the costs associated with fatalities and injuries caused by road traffic accidents in recent decades, considerable attention has been devoted to understanding the relationship between traffic accident severity and the influencing factors. This section provides a comprehensive review of current studies proposed in this context.

Jamal et al. investigated the effectiveness of the XGB-based model compared with traditional classifiers (LR, RF, and DT) for crash injury severity analysis. The data used in their research were obtained from motor vehicle collisions that occurred on rural highways in Riyadh, Saudi Arabia. Their experimental results demonstrated that the XGB classifier outperformed the traditional classifiers with a prediction accuracy of 93% [2].

To predict the severity level of accidents, Kumeda et al. applied various classifiers, including Fuzzy-FARCHD and RF, among others. Utilizing a dataset obtained from UK RTAs that occurred in 2016, their experimental results demonstrated the effectiveness of the Fuzzy-FARCHD algorithm in classifying severity levels, achieving an accuracy of 85.94% [3]. Using the same dataset, Gan et al. employed the novel Deep Forests algorithm, introduced for the first time by Zhou and Feng [4]. To validate their approach in terms of classification performance, various ML algorithms were implemented. Results indicate that the Deep Forests algorithm demonstrates stability, fewer hyperparameters, and superior accuracy across different levels of training data [5]. Al-Mamlook et al. (2019) compared various ML algorithms for predicting crash severity. The test results, with an AUC of 0.826, showcase the superior classification performance of RF in comparison with other classifiers [6].

To classify RTAs' severity, Labib et al. (2019) conducted an in-depth analysis of traffic accidents in Bangladesh using diverse ML algorithms. To simplify their analysis in terms of feature space dimensions, they applied three FS techniques: Univariate FS (UFS), Recursive Feature Elimination (RFE), and Feature Importance (FI). This resulted in a classification performance not exceeding 80% in terms of accuracy with AdaBoost, involving the use of 11 features for their analysis [7].

In their work, Chen et al. focused on addressing the impact of severe traffic accidents in China by analyzing complex relationships among contributing factors using a Bayesian network (BN) crash severity model. The limited data in terms of samples has impacted the prediction performance of their BN model [8].

In their work, Li et al. utilized an updated version of the UK RTAs dataset from 2017. They introduced a hybrid algorithm, namely Light Gradient Boosting Machine-Tree-structured Parzen Estimator (LightGBM-TPE). Consequently, their findings revealed that "Longitude," "Latitude," "Hour," and "Day_of_Week" are the four risk factors most strongly correlated with accident severity [9].

3 Methodology

In this section, we provide a comprehensive description of our proposed predictive approach. As illustrated in Fig. 1, our proposed approach consists of three main stages: data preprocessing, feature selection, and accident severity classification. More details about these stages are given in the subsections below.

3.1 Data Description and Preprocessing Phase

Data Description

The dataset proposed in this study was collected from the police departments of Addis Ababa Sub City for a Master's research project. The dataset, sourced from an open data platform [10], has been prepared from manual records of RTAs that occurred between the years 2017 and 2020. The dataset proposed for analysis includes 32 features and 12,316 instances of accidents, which are categorized into three classes: 10,415 instances labeled as Slight Injury, 1,743 instances labeled as Serious Injury, and 158 instances labeled as Fatal Injury. The input features description and their numerical assignment are summarized below.

1. **Time**: Indicates the time of the accident in hours and minutes.

2. **Day of Week**: Numerical representation of the day when the accident occurred (Monday to Sunday).
3. **Age Band of Driver**: Categorizes the age range of the driver (under 18, 18–30, 31–50, over 51).
4. **Sex of Driver**: Gender of the driver (male unknown, or female).
5. **Educational Level**: Educational level of the driver (illiterate, elementary school, high school, junior high school, above high school, writing & reading).

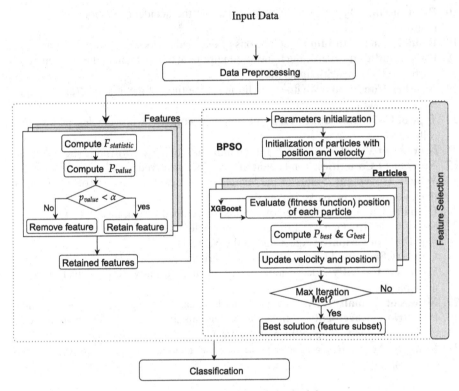

Fig. 1. The proposed approach methodology.

6. **Vehicle Driver Relation**: Association of the driver with the vehicle (owner: 0, employee: 1, other: 2).
7. **Driving Experience**: Level of driving experience (no license, below 1 year, 1–2 years, 2–5 years, 5–10 years, above 10 years).
8. **Type of Vehicle**: Type of vehicle involved (various categories, e.g., motorcycle, taxi, lorry).
9. **Owner of Vehicle**: Ownership of the vehicle (owner, governmental, organization, or other).
10. **Service Year of Vehicle**: Number of years the vehicle has been in service (below 1yr, 1–2yr, 2–5yrs, 5–10yrs, above 10yr).

11. **Defect of Vehicle**: Describes defects or issues with the vehicle involved (numeric).
12. **Area Accident Occurred**: Area where the accident occurred (e.g., rural village areas, market areas).
13. **Lanes or Medians**: Configuration of lanes or medians on the road (e.g., undivided, one-way, divided with broken lines).
14. **Road Alignment**: Alignment characteristics of the road at the accident location (e.g., tangent road with flat terrain, steep grade upward).
15. **Types of Junctions**: Type of junction at the accident location (e.g., no junction, crossing, T shape).
16. **Road Surface Type**: Type of road surface at the accident location (e.g., asphalt roads, gravel roads).
17. **Road Surface Conditions**: Conditions of the road surface (e.g., dry, wet, snow).
18. **Light Conditions**: Prevailing lighting conditions at the moment of the accident (e.g., daylight, darkness with lights lit).
19. **Weather Conditions**: Weather conditions at the time of the accident (e.g., normal, raining, snowy).
20. **Type of Collision**: Nature of the collision (e.g., collision with vehicles, pedestrians, rollover).
21. **Number of Vehicles Involved**: Total number of vehicles involved in the accident.
22. **Number of Casualties**: Total count of casualties involved.
23. **Vehicle Movement**: Status of the vehicle's movement at the time of the accident (e.g., going straight, parked, overtaking).
24. **Casualty Class**: Classification of victims (driver, pedestrian, passenger).
25. **Sex of Casualty**: Gender of the casualties (male, unknown, or female).
26. **Age Band of Casualty**: Age range of the casualties.
27. **Casualty Severity**: Degree of casualty severity.
28. **Work of Casualty**: Occupation status of the casualty (e.g., driver, student, unemployed).
29. **Fitness of Casualty**: Physical condition of the casualty (e.g., normal, deaf, blind).
30. **Pedestrian Movement**: Actions of individuals involved in the accident (e.g., crossing, walking along in the carriageway).
31. **Cause of Accident**: Categorized cause of the accident (e.g., overspeed, improper parking).

Data Preprocessing

Before subjecting data to any FS or classification process, it is imperative to conduct preprocessing on the raw data. Our preprocessing step involves data cleaning, where whitespaces are systematically removed, and anomalies are corrected. Data encoding, wherein we encode variables considering their individual characteristics. Data imputation, where we adopted a machine learning-based approach to address incomplete features. Finally, features are standardized using the Z-score Eq. (1), transforming them to have a mean of zero and a standard deviation (std) of one.

$$z_i = \frac{(x_i - \bar{x})}{S} \qquad (1)$$

Where z_i denotes the individual feature value, $\bar{x}$ is the mean of the feature, and S is its standard deviation.

3.2 Feature Selection Strategy

FS plays a crucial role in modeling intelligent systems, particularly in predictive systems [11–14]. The identification of the most relevant features in classification tasks offers numerous advantages, such as improved prediction performance, enhanced model interpretability, and cost reduction in terms of time and resources [15]. Commonly, FS methods are categorized as filters and wrappers [16]. In filter, the relevance of attributes is based on certain statistical criteria [17], while wrapper methods evaluate the relevance of feature subset by training a specific classification model [16]. Therefore, wrapper methods usually achieve better classification performance at the expense of increased time consumption. In this study, we conducted FS by proposing a two-step process that combines the One-Way-ANOVA-based filter method with a Binary Particle Swarm Optimization-based wrapper method.

One-Way-ANOVA-Based Filter Method
As an interesting technique, the one-way ANOVA examines variations in means among different groups (each group representing a class) within a dataset, facilitating the identification of statistically significant differences in feature values across groups. In the context of our approach, the relevance of each feature is assessed using a One-Way-ANOVA-based p-value. The key steps of the suggested method are outlined as follows:

For each feature X in the dataset, where n is the number of samples, k is the number of classes, n_i is the number of samples in class i, X_{ij} is the j-th samples in class i, $\overline{X}$ is the overall mean of the feature X, $df_{bc} = k - 1$ the degrees of freedom between classes, and $df_{wc} = n - 1$ the degrees of freedom within classes.

1. Compute the variance between classes:

$$M_{bc} \leftarrow \frac{\sum_{i=1}^{k} n_i (\overline{X_i} - \overline{X})^2}{df_{bc}}$$

2. Compute the variance within classes:

$$M_{wc} \leftarrow \frac{\sum_{i=1}^{k} \sum_{j=1}^{n_i} (X_{ij} - \overline{X_i})^2}{df_{wc}}$$

3. Compute the F-statistic:

$$F_{statistic} \leftarrow \frac{M_{bc}}{M_{wc}}$$

4. Find p-value based on the F-statistic: $p - value \leftarrow Prob(F > F_{statistic} | df_{bc}, df_{wc})$, where F is a random variable that follows the F-distribution.
5. If p-value < 0.05, then retain the feature.

Binary Particle Swarm Optimization-Based Wrapper Method
Developed by Kennedy and Eberhart [18], PSO is population-based metaheuristic technique designed to emulate social behavior observed in natural, such as the collective

movement of particles, akin to the coordination seen in fish schools or bird flocking. In PSO, a population, often referred to as a swarm, consists of candidate solutions represented as particles within a defined search space. Initially, PSO starts with a random placement of particles in a p-dimensional search space. Each particle i is characterized by its position $X_i = (x_{i1}, x_{i2}, \ldots, x_{ip})$, its velocity $V_i = (v_{i1}, v_{i2}, \ldots, v_{ip})$, and its personal best position $pbest_i = (pbest_{i1}, pbest_{i2}, \ldots, pbest_{ip})$. The entire swarm explores the search space to find the optimal solution by updating the position of each particle, which is influenced by its *pbest* and the best position of the swarm, denoted by $gbest = (gbest_1, gbest_2, \ldots, gbest_p)$. The assessment of *pbest* and *gbest* depends on a fitness function that quantifies the effectiveness of each particle's solution by evaluating its performance level. The velocity and the position of the particle are updated according the following equations:

$$v_{ij}(t+1) = w \cdot v_{ij}(t) + c1 \cdot r1 \cdot \left(pbest_{ij} - x_{ij}(t)\right) + c2 \cdot r2 \cdot \left(gbest_j - x_{ij}(t)\right) \quad (2)$$

$$x_{ij}(t+1) = x_{ij}(t) + v_{ij}(t+1) \quad (3)$$

Where $x_{ij}(t+1)$ and $v_{ij}(t+1)$ represent the new position and velocity of the particle *i* at the index *j* during iteration $(t+1)$. Meanwhile, $x_{ij}(t)$ and $v_{ij}(t)$ represent the current position and velocity of the particle *i*, respectively. *c1* and *c2* are the acceleration coefficients, *w* is an inertia weight used to control the global exploration and local exploitation of search area, and *r1* and *r2* are two random numbers uniformly distributed in [0, 1].

In the context of our FS approach, we have opted to use Binary Particle Swarm Optimization (BPSO), a variant introduced by Kennedy and Eberhart, specifically designed for addressing discrete problems [19]. As an efficient FS method [20–22], BPSO becomes particularly advantageous when making binary decisions about the inclusion or exclusion of features. In BPSO, a particle explores a discrete search space by using its velocity to determine the probability of each feature taking a value of one or zero. To restrict the continuous position values to a binary representation (0 or 1), the sigmoid function is employed, expressed in Eq. (4):

$$\text{Sig}(v_{ij}) = \frac{1}{1 + e^{-v_{ij}}} \quad (4)$$

This technique guarantees that, depending on the comparison between $\text{Sig}(v_{ij})$ and the randomly generated number, the particle's position x_{ij} is decisively assigned either the value of 1 or 0 (Eq. (5)).

$$x_{ij} = \begin{cases} 1 & \text{if } random(0, 1) < \text{Sig}(v_{ij}) \\ 0 & \text{otherwise,} \end{cases} \quad (5)$$

The specific steps of our BPSO-based wrapper approach are as follows:

Step 1: Initializing algorithm parameters.: *c1, c2, w, n* (number of particles), max number of iteration (it_{max}).
Step 2: Assigning random positions and velocities to the particles within the swarm.

Step 3: Evaluate the quality of each particle's generated feature subset (solution). In this process, the XGBoost classifier is employed to assess the effectiveness of the feature subsets, with accuracy serving as the chosen evaluation metric. The hold-out method is adopted as the evaluation technique.
Step 4: update the best personal solution (*pbest*) and the best global solution (*gbest*).
Step 5: Update the position and the velocity of each particle according to Eqs. (2)–(5).
Step 6: Repeat steps 3 to 5 until the maximum iteration limit (it_{max}) is met.
Step 7: The best feature subset is retained.

3.3 Classification

To assess the impact of our FS approach on the classification of accidents in terms of their severity levels, we diversified the ensemble of recommended machine learning algorithms. This diversification was achieved by incorporating a blend of classical and boosting algorithms, each category offering distinct advantages. Among the classical classifiers, KNN operates on the principle of proximity, assigning labels based on the majority class among neighboring data points [23]. DT divide the data into subsets based on feature criteria, forming a tree-like structure [24]. RF is an enhanced version of DT that uses a collection of decision trees to improve predictive accuracy and mitigate overfitting [25]. ANN mimic the human brain's structure, featuring interconnected nodes and layers for complex pattern recognition [26]. In the realm of boosting algorithms, LGBM sequentially builds weak models, correcting errors at each step [27], while XGB employs optimization techniques for efficient and accurate predictions [28]. This diverse set of classifiers allows for a nuanced examination of our FS strategy's efficacy across a spectrum of machine learning methodologies.

4 Experimental Results

4.1 Validation Techniques and Evaluation Metrics

In this study, we applied the stratified K-fold cross-validation (CV) technique. Through this method, the dataset is shuffled and split into 10 folds. Nine folds are used for model construction, with the remaining fold reserved for testing the constructed models. This technique is employed to guarantee a balanced representation of classes during both model training and testing, which can enhance the robustness of our evaluation methodology, providing a stable and reliable estimate of the model's performance.

Based on existing literature, the classification performance of the suggested machine learning algorithms was assessed using evaluation metrics such as accuracy, precision, recall, and F-score (Eq. (6)–(9)). These metrics provide a comprehensive understanding of the models' effectiveness and capturing the balance between true positives, false positives, and false negatives.

$$\text{Accuracy} = \frac{\text{TP} + \text{TN}}{\text{TP} + \text{FP} + \text{TN} + \text{FN}} \tag{6}$$

$$\text{Recall} = \frac{\text{TP}}{\text{TP} + \text{FN}} \tag{7}$$

$$\text{Precision} = \frac{\text{TP}}{\text{TP} + \text{FP}} \qquad (8)$$

$$\text{F1} - \text{score} = 2 \times \frac{\text{Recall} \times \textit{Precision}}{\text{Recall} + \textit{Precision}} \qquad (9)$$

TP (True Positive) represents correctly identified positive instances, FP (False Positive) signifies wrongly identified positive instances among the actual negatives, TN (True Negative) denotes accurately predicted instances of a negative class, and FN (False Negative) indicates inaccurately predicted negative instances among the actual positives.

As we are addressing a multi-class problem, all proposed metrics are computed using the weighted average technique. This technique is particularly used to deal with imbalanced distributions of instances across different classes, as is the case with our dataset. The weighted average method ensures that the impact of each class on the overall metric calculation is proportionate to its representation in the dataset (Eq. (10)).

$$\text{Weighted\_Average\_Metric} = \frac{\sum_{i=1}^{n} w_i \cdot \text{Metric}_i}{\sum_{i=1}^{n} w_i} \qquad (10)$$

Where n represent the number of classes and w_i denotes the weight assigned to the class i.

4.2 Parameter Setting

The entire proposed severity prediction system was developed in Python 3.7 and tested on a Windows 10 Pro for Workstations, equipped an Intel Xeon processor (with a frequency of 3.5 GHz) and 64 GB of RAM. To ensure consistency in comparison, the feature selection process outlined in this study utilized predefined parameter settings for our Binary Particle Swarm Optimization BPSO-based wrapper method within the same runtime environment. Specifically, the algorithm was configured with the following parameters: the number of particles (n) set to 20, the maximum number of iterations (it_{max}) set to 100, c1 and c2 assigned values of 1.0 and 2.5, respectively, w (inertia weight) set at 0.8, and r1 and r2 representing two random numbers uniformly distributed in the range [0, 1].

4.3 Results and Discussion

To analyze the impact of our proposed FS approach, OWABPSO, on RTA severity prediction, this section discusses and presents the results of the six suggested classifiers—DT, ANN, kNN, LGBM, RF, and XGBoost—in terms of classification performance. The outcomes are examined across both the entire feature space and a reduced space achieved through our OWABPSO approach.

Performance Analysis Without Feature Selection

The Table 1 illustrates the classification performance metrics, in terms of accuracy, precision, recall, and F1-score, with values provided for mean, max, min, and standard deviation (std) for each classifier across the entire feature space (without using any dimensionality reduction approach). As we can notice, the XGBoost stands out as the top-performing model with the highest classification performance in terms of all evaluation metrics, accuracy (85.97%), precision (83.95%), recall (85.97%), and F1-score (81.98%), showcasing its robust classification capabilities using all the feature space. On the other hand, DT-based model exhibited comparatively lower performance, indicating potential challenges in achieving a balanced trade-off between precision and recall. Generally, the remaining classifiers achieve classification performance that hovers around 84% in terms of accuracy.

Table 1. Classification performance with all the feature space for different classifiers

Model	Accuracy (%)	Precision (%)	Recall (%)	F1-score (%)
	\multicolumn{4}{c}{$(Mean_{Min}^{Max} \mp std)$}			
DT	$(77.43_{76.06}^{79.14} \mp 0.84)$	$(78.81_{77.30}^{80.08} \mp 0.96)$	$(77.43_{76.06}^{79.14} \mp 0.84)$	$(78.06_{77.10}^{79.45} \mp 0.77)$
ANN	$(84.57_{84.50}^{84.58} \mp 0.03)$	$(71.53_{71.52}^{71.53} \mp 0.00)$	$(84.57_{84.50}^{84.58} \mp 0.03)$	$(77.51_{77.47}^{77.51} \mp 0.01)$
KNN	$(84.50_{84.25}^{84.66} \mp 0.13)$	$(75.15_{71.50}^{85.72} \mp 4.75)$	$(84.50_{84.25}^{84.66} \mp 0.13)$	$(77.57_{77.38}^{77.82} \mp 0.14)$
LGBM	$(85.69_{85.23}^{86.04} \mp 0.28)$	$(84.47_{82.88}^{85.66} \mp 0.93)$	$(85.69_{85.23}^{86.04} \mp 0.28)$	$(80.71_{79.61}^{81.35} \mp 0.55)$
RF	$(85.24_{84.98}^{85.47} \mp 0.15)$	$(85.39_{84.33}^{86.30} \mp 0.74)$	$(85.24_{84.98}^{85.47} \mp 0.15)$	$(79.17_{78.61}^{79.57} \mp 0.29)$
XGB	$(85.97_{85.23}^{86.53} \mp 0.42)$	$(83.95_{81.81}^{85.05} \mp 1.21)$	$(85.97_{85.23}^{86.53} \mp 0.42)$	$(81.98_{81.04}^{82.88} \mp 0.55)$

Performance Analysis with Feature Selection

Table 2 reports experimental results obtained using our proposed FS approach, which means working with a feature space of only two features: hour and minute. As observed, in comparison with Table 1 where no dimensionality reduction approach was applied, the suggested classifiers could maintain robust performance even with a reduced feature space. For instance (also refer to Fig. 2), the ANN classifier maintains an accuracy of approximately 84.58% both without and with FS. A similar pattern is observed for the other classifiers, such as DT, KNN, LGBM, RF, and XGB, where the performance remains consistent or even slightly improved with feature selection. For instance, the precision of the DT classifier increases from 78.81% to 81.24% when applying our FS approach. Similarly, the F1-score for the XGB classifier improves from 81.98% to 79.99%.

Figure 3 illustrates the ROC curves, depicting the balance between correctly identifying instances of severe road traffic accidents (sensitivity), and the misclassification of less severe accidents, denoted by 1-specificity. In this context, where three classes are considered—class 0 for Slight Injury, class 1 for Serious Injury, and class 2 for Fatal Injury—the figure provide a visual representation of the models' ability to discern

Table 2. Classification performance with reduced feature space for different classifiers

Model	Accuracy (%)	Precision (%)	Recall (%)	F1-score (%)
	(Mean $_{Min}^{Max}$ ∓ std)			
DT	(84.90$_{84.50}^{85.88}$ ∓ 0.41)	(81.24$_{80.19}^{83.99}$ ∓ 1.15)	(84.90$_{84.50}^{85.88}$ ∓ 0.41)	(80.86$_{80.27}^{82.26}$ ∓ 0.61)
ANN	(84.58$_{84.58}^{84.58}$ ∓ 0.00)	(71.53$_{71.53}^{71.53}$ ∓ 0.00)	(84.58$_{84.58}^{84.58}$ ∓ 0.00)	(77.51$_{77.51}^{77.51}$ ∓ 0.00)
KNN	(84.41$_{84.17}^{84.66}$ ∓ 0.14)	(78.31$_{76.07}^{79.81}$ ∓ 1.23)	(84.41$_{84.17}^{84.66}$ ∓ 0.14)	(78.57$_{77.79}^{79.14}$ ∓ 0.41)
LGBM	(84.86$_{84.50}^{85.23}$ ∓ 0.22)	(81.43$_{78.71}^{83.44}$ ∓ 1.82)	(84.86$_{84.50}^{85.23}$ ∓ 0.22)	(78.90$_{78.32}^{79.84}$ ∓ 0.42)
RF	(84.63$_{83.60}^{85.39}$ ∓ 0.50)	(80.69$_{78.60}^{82.58}$ ∓ 1.18)	(84.63$_{83.60}^{85.39}$ ∓ 0.50)	(80.83$_{79.69}^{82.05}$ ∓ 0.70)
XGB	(84.98$_{84.25}^{85.71}$ ∓ 0.45)	(81.49$_{78.86}^{83.88}$ ∓ 1.56)	(84.98$_{84.25}^{85.71}$ ∓ 0.45)	(80.00$_{79.26}^{81.30}$ ∓ 0.59)

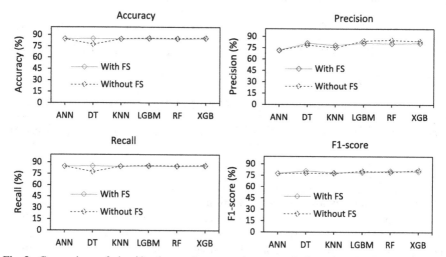

Fig. 2. Comparison of classification performance, in terms of all used evaluation metrics, for suggested classifiers with and without feature selection.

between these severity levels. As observed, with the exception of the ANN-based model, all decision tree-based models indicate that our FS approach had minimal impact on their ability to predict accident severity.

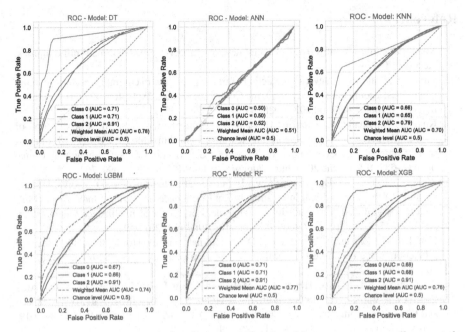

Fig. 3. ROC cure for classification performance with reduced feature space across all suggested classifiers (with class 0: Slight Injury, class 1: Serious Injury, class 2: Fatal injury)

5 Conclusion

The prediction of traffic accident severity is a crucial research topic for the implementation of advanced road safety strategies. This work aimed to propose a new accident severity prediction strategy that consists in integrating a hybrid FS approach, OWABPSO, with machine learning algorithms. The OWABPSO comprises a two-step process, combining the One-Way-ANOVA-based filter method with a Binary Particle Swarm Optimization-based wrapper method. The results showed that by involving only time-based variables (Hours and minutes) using our FS approach, we could maintain almost the same classification performance as obtained without using any dimensionality reduction method on the proposed dataset, sometimes even with a slight improvement in performance, particularly for the Decision Tree classifier. Generally, all of classifiers exhibited acceptable performance, not falling below 84.5% in terms of accuracy. This suggests the potential utility of dimensionality reduction approaches in enhancing accident severity prediction performance, thereby aiding traffic safety engineers in implementing effective road safety strategies.

Based on this study, future work involves using other datasets to evaluate our proposed approach further. Additionally, to enhance the credibility of our study, we intend to conduct a more in-depth analysis, both on the proposed datasets and the obtained results. Furthermore, we aim to propose an improved version of our FS approach, refining its performance and addressing potential limitations identified in this work.

References

1. Road traffic injuries. https://www.who.int/news-room/fact-sheets/detail/road-traffic-injuries. Accessed 19 Nov 2023
2. Jamal, A., et al.: Injury severity prediction of traffic crashes with ensemble machine learning techniques: a comparative study. Int. J. Inj. Contr. Saf. Promot. **28**(4), 408–427 (2021). https://doi.org/10.1080/17457300.2021.1928233
3. Kumeda, B., Zhang, F., Zhou, F., Hussain, S., Almasri, A., Assefa, M.: Classification of road traffic accident data using machine learning algorithms. In: 2019 IEEE 11th International Conference on Communication Software and Networks (ICCSN), Chongqing, China, pp. 682–687. IEEE, June 2019. https://doi.org/10.1109/ICCSN.2019.8905362
4. Zhou, Z.-H., Feng, J.: Deep forest: towards an alternative to deep neural networks. In: Proceedings of the Twenty-Sixth International Joint Conference on Artificial Intelligence, Melbourne, Australia: International Joint Conferences on Artificial Intelligence Organization, pp. 3553–3559, August 2017. https://doi.org/10.24963/ijcai.2017/497
5. Gan, J., Li, L., Zhang, D., Yi, Z., Xiang, Q.: An alternative method for traffic accident severity prediction: using deep forests algorithm. J. Adv. Transp. **2020**, 1–13 (2020). https://doi.org/10.1155/2020/1257627
6. Al Mamlook, R.E., Ali, A., Hasan, R.A., Mohamed Kazim, H.A.: Machine learning to predict the freeway traffic accidents-based driving simulation. In: 2019 IEEE National Aerospace and Electronics Conference (NAECON), Dayton, OH, USA, pp. 630–634. IEEE, July 2019. https://doi.org/10.1109/NAECON46414.2019.9058268
7. Labib, Md.F., Rifat, A.S., Hossain, Md.M., Das, A.K., Nawrine, F.: Road accident analysis and prediction of accident severity by using machine learning in Bangladesh. In: 2019 7th International Conference on Smart Computing & Communications (ICSCC), Sarawak, Malaysia, Malaysia, pp. 1–5. IEEE, June 2019. https://doi.org/10.1109/ICSCC.2019.8843640
8. Chen, H., Zhao, Y., Ma, X.: Critical factors analysis of severe traffic accidents based on Bayesian network in China. J. Adv. Transp. **2020**, e8878265 (2020). https://doi.org/10.1155/2020/8878265
9. Li, K., Xu, H., Liu, X.: Analysis and visualization of accidents severity based on LightGBM-TPE. Chaos Solitons Fractals **157**, 111987 (2022). https://doi.org/10.1016/j.chaos.2022.111987
10. Bedane, T.T.: Road traffic accident dataset of addis ababa city [Object], 02 November 2020. https://doi.org/10.17632/XYTV86278F.1
11. Hamim, M., El Moudden, I., Pant, M.D., Moutachaouik, H., Hain, M.: A hybrid gene selection strategy based on fisher and ant colony optimization algorithm for breast cancer classification. Int. J. Online Eng. **17**(02), 148 (2021). https://doi.org/10.3991/ijoe.v17i02.19889
12. Hamim, M., El Mouden, I., Ouzir, M., Moutachaouik, H., Hain, M.: A novel dimensionality reduction approach to improve microarray data classification. IIUMEJ **22**(1), 1–22 (2021). https://doi.org/10.31436/iiumej.v22i1.1447
13. El Moudden, I., Ouzir, M., ElBernoussi, S.: Feature selection and extraction for class prediction in dysphonia measures analysis: a case study on Parkinson's disease speech rehabilitation. Technol. Health Care off. J. Eur. Soc. Eng. Med. **25**, 1–16 (2017). https://doi.org/10.3233/THC-170824
14. Moutachaouik, H., El Moudden, I.: Mining prostate cancer behavior using parsimonious factors and shrinkage methods (2018)
15. Chu, C., Hsu, A.-L., Chou, K.-H., Bandettini, P., Lin, C.: Does feature selection improve classification accuracy? Impact of sample size and feature selection on classification using anatomical magnetic resonance images. Neuroimage **60**(1), 59–70 (2012)

16. Kohavi, R., John, G.H.: Wrappers for feature subset selection. Artif. Intell. **97**(1), 273–324 (1997)
17. Guyon, I., Elisseeff, A.: An introduction to feature extraction. In: Guyon, I., Nikravesh, M., Gunn, S., Zadeh, L.A. (eds.) Feature Extraction. SFSC, vol. 207, pp. 1–25. Springer, Heidelberg (2006). https://doi.org/10.1007/978-3-540-35488-8_1
18. Kennedy, J., Eberhart, R.: Particle swarm optimization. In: Proceedings of ICNN'95 - International Conference on Neural Networks, vol. 4, pp. 1942–1948, November 1995. https://doi.org/10.1109/ICNN.1995.488968
19. Kennedy, J., Eberhart, R.C.: A discrete binary version of the particle swarm algorithm. In: Computational Cybernetics and Simulation 1997 IEEE International Conference on Systems, Man, and Cybernetics, vol. 5, pp. 4104–4108, October 1997. https://doi.org/10.1109/ICSMC.1997.637339
20. Wei, J., et al.: A BPSO-SVM algorithm based on memory renewal and enhanced mutation mechanisms for feature selection. Appl. Soft Comput. **58**, 176–192 (2017). https://doi.org/10.1016/j.asoc.2017.04.061
21. BinSaeedan, W., Alramlawi, S.: CS-BPSO: hybrid feature selection based on chi-square and binary PSO algorithm for Arabic email authorship analysis. Knowl. Based Syst. **227**, 107224 (2021). https://doi.org/10.1016/j.knosys.2021.107224
22. Kushwaha, N., Pant, M.: Link based BPSO for feature selection in big data text clustering. Future Gener. Comput. Syst. **82**, 190–199 (2018). https://doi.org/10.1016/j.future.2017.12.005
23. Cover, T., Hart, P.: Nearest neighbor pattern classification. IEEE Trans. Inform. Theory **13**(1), 21–27 (1967). https://doi.org/10.1109/TIT.1967.1053964
24. Hamim, M., El Moudden, I., Moutachaouik, H., Hain, M.: Decision tree model based gene selection and classification for breast cancer risk prediction. In: Hamlich, M., Bellatreche, L., Mondal, A., Ordonez, C. (eds.) SADASC 2020. CCIS, vol. 1207, pp. 165–177. Springer, Cham (2020). https://doi.org/10.1007/978-3-030-45183-7_12
25. Breiman, L.: Random Forests–Random Features, p. 29 (1999)
26. Cachim, P.: Using artificial neural networks for calculation of temperatures in timber under fire loading. Constr. Build. Mater. **25**, 4175–4180 (2011). https://doi.org/10.1016/j.conbuildmat.2011.04.054
27. Ke, G., et al.: LightGBM: a highly efficient gradient boosting decision tree. In: Advances in Neural Information Processing Systems. Curran Associates, Inc. (2017). Accessed 14 Dec 2023. https://proceedings.neurips.cc/paper_files/paper/2017/hash/6449f44a102fde848669bdd9eb6b76fa-Abstract.html
28. Ibrahem Ahmed Osman, A., Najah Ahmed, A., Chow, M.F., Feng Huang, Y., El-Shafie, A.: Extreme gradient boosting (Xgboost) model to predict the groundwater levels in Selangor Malaysia. Ain Shams Eng. J. **12**(2), 1545–1556 (2021). https://doi.org/10.1016/j.asej.2020.11.011

The Integration of NLP and Topic-Modeling-Based Machine Learning Approaches for Arabic Mobile App Review Classification

Daniel Voskergian[1](✉) and Faisal Khamayseh[2]

[1] Computer Engineering Department, Al-Quds University, Jerusalem, Palestine
daniel2vosk@gmail.com
[2] Computer Science Department, Polytechnic University, Hebron, Palestine
faisal@ppu.edu

Abstract. App stores serve as digital distribution platforms for mobile applications for almost every software and service. They provide users with a range of applications to browse, purchase or install at no cost. Additionally, users can post reviews in terms of textual feedback and star rating. According to recent studies, text reviews provide rich information, such as enhancement requests, bug reports, user experience, and text ratings. Such data can benefit different stakeholders, especially developers, who can examine the customer needs and react through app improvement, thus increasing the app's popularity, quality and success in the marketplace. However, the amount of app reviews is too sheer for manual categorization. Thus, an automated approach is required to support developers in analyzing app reviews. While many studies have proposed approaches using classical machine learning algorithms for English text reviews, there is limited research on applying Arabic Text Classification for user reviews in Arabic. This paper aims to leverage Arabic app reviews, natural language processing and topic modeling-based machine learning techniques to classify and analyze user feedback.

Keywords: App Reviews · Arabic Text Classification · Machine Learning · Natural Language Processing · Topic Modeling

1 Introduction

Traditionally, requirement engineering (RE) has been driven by stakeholders, where the requirements are elicited from domain knowledge via qualitative data collection methods (ex., workshops, interviews, observations, and focus group discussions) [1]. These conventional techniques are often time-consuming and are not capable of handling stakeholder groups that are expanding in both size and global reach. However, with the widespread digitization we are witnessing globally, vast amounts of data are generated daily from various digital sources (ex., social media and app stores). These dynamic data (e.g., microblogs, online reviews, etc.) can be considered a valuable source of requirements, even though they are not intentionally created for requirements elicitation.

Recently, a dominating approach called Crowd-based requirements engineering (CrowdRE) has taken its place in the field of RE. Its primary focus has been eliciting requirements from explicit user feedback (ex., app reviews, and social media data) by applying various natural language processing techniques and machine learning algorithms [2].

We can gain benefits from eliciting requirements from dynamic data created by digital sources. First, there is no need to collect data specifically for RE since it facilitates secondary and ready-to-use data. Second, digital data sources allow us to collect data relevant to new system requirements obtained from global stakeholders beyond an organization's reach. Adding those requirements to the system can bring business value through improved customer satisfaction and optimized operation. Third, dynamic data allows for collecting up-to-data user requirements, which can be utilized in timely and effective operational decision-making [3, 4].

To this end, numerous efforts were placed to automate requirement elicitation from static domain knowledge that is rarely updated, i.e., documents written in natural language and different types of models (e.g., UML use case and sequence diagrams). However, eliciting requirements from dynamic data that are not intentionally created for this purpose has received much less effort.

Nowadays, app reviews collected by various app stores are gaining attention in Requirement Engineering. App stores serve as digital distribution platforms for desktop and mobile applications. They provide users with various applications to browse through, purchase, or install at no cost. Moreover, users have the option to provide reviews through textual feedback and star ratings. According to the latest estimates, 257 billion apps were installed worldwide in 2022, of which 143 billion were installed from the top two largest app stores, Apple App Store and Google Play [5].

In the app distribution platforms, users commonly share their app experience and interaction feedback by posting reviews for the installed applications. These reviews, displayed on various websites and platforms, serve to establish an online presence and foster trust among potential customers [3]. According to [6], the top ten apps they studied receive an average of 500 reviews per day, and for top-rated apps, this number can escalate to several thousand reviews daily.

Recent research indicates that user reviews on these platforms offer valuable information that can benefit various stakeholders, including app vendors and developers. These reviews encompass user sentiments regarding specific features, identification of feature bugs, suggestions for improvements, proposals for new features, documentation of released features, and detailed descriptions of user experiences, opinions, and concerns related to different features. It is worth noting that some reviews may also contain irrelevant information, such as advertisements, spam, and star ratings expressed in words [7, 8].

Such insightful information allows developers to examine customer needs and react through app improvement. This process will ultimately enhance the app's popularity, ranking, quality, and market success. However, if developers do not recognize and act upon such information on time, app users could simply install and use an alternative app, given the number of choices available. In addition, new developers or senior developers working on a new app can benefit from reviews that users post on other similar apps in the same domain, as they can extract feature requests, associate them with real functionality, and integrate them inside the initial release of their app. Thus, they can deliver their app in a shorter period, enriched with trending features that users most prefer.

However, the amount of app reviews is too sheer for manual processing. It is incredibly time-consuming and labor-intensive for developers to go through them one by one to identify those reviews that are useful for their purposes. Moreover, their quality may vary to a large extent when we consider the amount of noise and ambiguities inherent to natural language. Thus, an automated approach is required to support developers in analyzing app reviews.

To this end, several automated approaches have been proposed for the automatic classification of reviews, employing classical machine learning algorithms. Among these studies, the majority used textual information written in English, with the minority focusing on other languages, such as Arabic.

This paper investigates the field of research using Arabic Text Classification (ATC) on mobile app reviews to extract requirement-related information. As reported by the literature, there is a scarcity of studies exploring ATC for the mobile app requirements evolution. Leveraging ATC approaches and tools holds the potential to aid developers in advancing the software evolution of mobile apps. The study utilizes app reviews written in Arabic as a valuable source of requirements.

Topic models (TM) are unsupervised machine-learning algorithms that uncover latent semantic structures within extensive text documents. Conceived initially as a text-mining technique for revealing meaningful hidden patterns and interpretable semantic concepts, they have since found application in text classification. In this context, the topics extracted from a vast text collection serve as features for document representation [9].

In this aspect, we utilized previously developed topic modeling-based text classification algorithms called TextNetTopics [10] and TextNetTopics Pro [11]. TextNetTopics aims to find the top r topics extracted by a topic model to train the classifier, where a topic is a group of semantically related words [10]. In contrast, TextNetTopics Pro is an advanced version tailored for short text. It seeks to uncover a promising combination of topics comprising words and topic distribution extracted by a short text topic model, serving as the basis for training a classifier [11].

Both approaches are employed to develop multiple binary classifiers customized for a multi-class classification problem. The primary goal is categorizing requirement-related information embedded within Arabic app reviews into distinct classes: A) *Enhancement Requests*: Users propose requests for additional functionality and content or share ideas for improving the app. B) *Bug Reports*: Users report system bugs, including crashes, incorrect behavior, or performance issues that need attention. C) *User Experience*: Users express sentiments, preferences, perceptions, psychological responses, and achievements related to their interaction with the app or specific features. This includes feedback before, during, and after usage. D) *Text Rating*: Users provide a numerical rating within a specified scale (e.g., from 1 to 10) accompanied by written text summarizing their overall assessment and feedback on the app.

The remainder of this paper is organized as follows: Sect. 2 delves into previous related work, exploring the existing literature in the field. Section 3 outlines our proposed methodology, detailing the approaches and techniques employed in our study. Subsequently, Sect. 4 offers insights into the experimental setup and presents the evaluation results. Finally, Sect. 5 concludes the paper, summarizing key findings, and discusses potential avenues for future research.

2 Related Work

App review classification in the English Language has been examined by several studies [12–16]. Four of these research papers [13–16] utilize a machine learning approach, while [12] use the deep learning-based model to classify mobile app reviews.

In terms of the classification classes, the mentioned studies had several differences in determining the classes of app reviews. [12, 14] used four classes: bug report, feature request, user experience, and ratings; although [12] used the term enhancement feature, it has the same meaning as the feature request. [13] used five classes: feature request, problem discovery, information giving, information seeking, and others, while the author in [15] utilized only the first three classes from [13]. In this aspect, the problem discovery class in [13, 15] has the same meaning as a bug or bug report in [12, 14, 16]. However, in [16], they used nine classes divided into two levels. The first level classifies reviews into a bug, new feature request, or sentiment. The second level, the bug review, is further classified into general, performance, security, and usability bugs. In contrast, the sentiment review is classified into positive, negative, or neutral. We can notice that these five studies agree on two essential app reviews: feature requests and bug reports.

Those studies used different datasets they created due to the lack of a standard dataset. [14] created an unbalanced dataset of 4,400 text reviews with multi-label classes. In [12], the author used the same dataset created in [14] in their research. [15] down-sampled reviews from 1,472 sentences to 438 single-labeled sentences to create a balanced dataset, with 146 sentences for each class. The study in [16] used 7,500 reviews that were single-labeled and unbalanced, and [13] used 1,390 reviews that were also single-labeled and unbalanced. In this aspect, [12–14] used Google Play Store and App Store as the primary source for collecting app reviews, while [15] and [16] used only Google Play Store. For the labeling process, studies [14–16] used expert annotators to manually label each review in the dataset, while the work in [12] used text matching using SQL queries.

Concerning the preprocessing techniques, most studies used natural language processing tools. All five studies conducted stemming and stop-word removal in the preprocessing phase. [12, 14] used spell checking and lemmatization in each review. [12, 15] used tokenization, but at different levels; the authors in [13, 15] tokenize the sentences since they are classifying the reviews at sentence-level granularity, while the authors in [12] tokenize each review into words using Python Natural Language Toolkit (NLTK) to leverage word embedding. [15, 16] used text cleansing, i.e., removing noisy data (e.g., spaces, symbols, numbers). Reference [15] used negation handling, and reference [12] used lowercase conversion.

After preprocessing phase, feature extraction was performed. [12] used two groups of features, a bag of words and metadata: star rating, review length, tense detection, and one or two sentiment scores. [15] used the same metadata in [14] but also leveraged additional feature bigram in their study. [13] used unigram, linguistic pattern, and sentiment score between [−1, 1]. [16] used only unigram bigram. At the same time, reference [12] also exploited the non-textual information of each app review, such as the statistical data for the ap (e.g., reviewer statistics such as the total number of reviews posted by the reviewer, submission rate) and metadata associated with each app review.

The four works [13–16] compare the effectiveness of various supervised machine learning algorithms for classifying app review text. All previous studies but [14] perform experiments using Naïve Bayes and Decision Tree. References [13, 15, 16] also used Support Vector Machine (SVM), [14] used MaxEnt, and [13, 15] used Logistic Regression. However, the author [12] used a convolutional neural network (CNN) to perform their study.

As reported by the literature, there is a scarcity of studies exploring Arabic text classification (ATC) and Arabic sentiment analysis (ASA) on user text reviews of mobile apps for requirements elicitation and evolution. Saudy et al. confirm that this field has not been well-researched [17].

3 Proposed Methodology

To achieve the goal of this study, we aim to build a suitable classifier that can categorize each Arabic review into one or more of the predefined four classes: Enhancement Request, Bug Report, User Experience, and Text Rating. The classifier will be helpful for requirement engineering. The following subsections describe the six phases we will go through in this study. Figure 1 illustrates an overview of the proposed methodology.

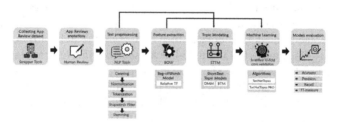

Fig. 1. Conceptual diagram of the proposed methodology.

3.1 Collecting App Review Dataset

To perform our study, we used Arabic app reviews for different apps within the same category to increase generalizability; precisely, we extracted reviews from three famous music streaming apps used in the Middle East (i.e., Anghami, SoundCloud, and Spotify). We utilized a Python-based package, a google-play-scrapper, to extract and collect app reviews from the Google Play Store. At the end of this step, we created a dataset of nearly 100,000 reviews that contains the following fields (user ID, published date, review text, app name, and app store name).

While other types of mobile apps, such as mHealth apps, represent a valuable domain to study, our research prioritized music apps due to their popularity and relevance in the Middle East. This ensured a substantial dataset of user reviews for analysis.

3.2 App Review Annotations

In the initial data collection phase, we took a random and representative sample of reviews for labeling purposes. Specifically, we chose 1,200 reviews from the Google Play Store. To ensure equal coverage for the three selected apps, we distributed these reviews into three distinct subsamples, each app encompassing 400 reviews.

For the labeling process, we hired two experts in the field (i.e., software engineering and computer science specialists) to manually annotate each review as one or more of four predefined classes: Enhancement Request, Bug Report, User Experience, and Text Rating. In this respect, an annotator has the ability to assign more than one category to a review. For instance, the review "الغنائية تطبيق يستحق خمسة نجوم ستجد كل ما تحتاجه اليه من الانواع", meaning "An application that deserves five stars for all the types of music you need " is classified as test rating (" تطبيق يستحق خمسة نجوم ") and a user experience (" كل ما تحتاجه اليه من الانواع الغنائية ستجد."). In this study, we split each review into multiple records because we emphasize multi-class classification rather than multi-label classification. This approach allows us to assign different categories to distinct review parts, providing a more detailed and nuanced understanding of the diverse aspects captured within each user's feedback. To simplify the labeling process, we build a Java-based user interface, that helps annotators concentrate on one review at a time.

Given the significant impact of miscategorization on the model's accuracy and the reliability of results, we have established a criterion for selecting labeled reviews. Specifically, we consider a review for model training only if both experts agree on the same category, thereby enhancing our confidence in the data quality. At the end of this phase, we successfully created a dataset consisting of 1,100 labeled reviews, ensuring a robust foundation for subsequent analyses and model development. Table 1 provides examples of manually annotated app reviews classified into four categories.

Table 1. Examples of annotated app reviews belong to the four categories.

Category	App Reviews
Enhancement Report	- الرجاء الغاء خدمه الكتروني وتسهيل الاستعمال وتحميل الفيديوهات والاغاني بدون ادخال البريد الالكتروني
Bug-Fix Report	- لا استطيع الدخول للتطبيق نهائيا من البارحة كل ما احاول ان ادخل اليه يخرج تلقائيا وبسرعة ارجو الحل.... خاصة واني في حالة اشتراك اسبوعية مع انغامي ... 💗 😢
User Experience	- تطبيق رائع لقد أحببته كثيرا،بمكنك الإستماع إلى أغانيك المفضلة في أي زمان ومكان 👀
Text Rating	- تطبيق يستحق اكثر من مليون نجمة 💗💗💗💗💗💗 The app deserves million stars

3.3 Data Preprocessing

App reviews manifest in natural language, carrying inherent noise, irrelevant details, and an unstructured format. Therefore, when preparing the input of app reviews into a classification model, a preprocessing step is imperative to transform the data into a clean and structured format. Similar to [18, 19], we utilized various natural language processing (NLP) techniques to prepare our textual data for subsequent analysis and processing. Proper application of these techniques can significantly impact the syntactical quality of the mining process output, contributing to enhanced accuracy in our classification models.

Cleaning. The cleaning phase is essential in our preprocessing pipeline. This step encompasses the removal of various forms of noisy data, including advertisements, spam, URLs, HTML tags, hashtags, numbers, repeating characters, punctuations, emojis, symbols and Non-Arabic content.

To execute this cleaning process, we employed regular expressions with the assistance of the Python re package. By systematically eliminating these elements, we aim to enhance the purity and relevance of the text data, preparing it for subsequent analysis and modeling.

Normalization. This step involves the conversion of non-standard words into their base forms. It includes the following processes: A) Diacritic Removal: Transforming words with diacritics to their diacritic-free equivalents (e.g., "أُرِيدُ" to "أريد," meaning "I want"). B) Elongation Removal: Resolving elongated forms to their base forms (e.g., "جمـــيل" to "جميل," meaning "beautiful"). C) Normalization of Special Letters: Standardizing special letters considered extras in the Arabic alphabet. This includes replacing "آ", "إ", and "أ" with "ا", converting "ؤ" and "ئ" to "ء", replacing "ة" with "ه", and changing the letter "ي" to "ى". This normalization practice is widely adopted in various research works within Arabic computational linguistics.

Tokenization. Tokenization is the process of segmenting review texts into a list of linguistic units or tokens, where each token can represent a paragraph, sentence, word, or character. In implementing this step, we employed the Python NLTK library, using non-letter (space) as the tokenization mode to split words in a review. Tokenizing this review ("ارجو تفعيل خاصية تنزل الأغاني على الهاتف المحمول"), meaning "Please activate the feature of downloading songs to our mobile phone", yields the result ([' ارجو', ' تفعيل', ' خاصية', ' تنزل', ' الأغاني', ' على', ' الهاتف', ' المحمول']).

Stemming. Stemming is a process of removing prefixes and suffixes from words, returning them to their root form [20]. For instance, stemming the word "المتطلبات" results in "متطلب", meaning "requirements". It's worth noting that stemming doesn't rely on any root dictionaries. In contrast, lemmatization employs dictionaries and considers the linguistic context to transform syntactically different tokens but semantically equivalent into their base form [3]. For example, both "متطلب" and "مطلب" can be lemmatized to "طلب", meaning "request". The process of associating variants with a root form increases the frequency of words used in reviews while reducing the overall dictionary size. To implement stemming in this study, we utilized the Arabic Snowball Stemmer node within the KNIME analytics platform.

Remove Arabic Stop Words. In this step, we aim to eliminate common tokens that carry minimal meaning, known as stop words., known as stop words. Stop words are frequently used terms that connect words within a sentence but contribute little to the overall meaning. Removing these stop words is beneficial for the classification process, significantly reducing noise in the data [21].

The NLP Python packages corpus comprises 248 Arabic stop words (e.g., 'التي', 'إذ', 'في', 'لك', 'إلى'). However, we will customize the stop word list to ensure that key keywords relevant to review classification (such as those indicating enhancement requests, bug reports, user experiences, and ratings) are preserved. This customization helps balance noise reduction and retaining essential information for the classification process.

3.4 Text Vectorization (Feature Extraction)

The vectorization process transforms unstructured free text into a structured numerical format, known as a vector of properties or a feature vector. Various techniques can represent each review text as a vector, such as Bag of Words models (e.g., TF or TF-IDF) and Word embeddings (e.g., Word2vec or Glove). In this study, we opt for the use of relative term frequency. This measure is derived by considering the frequency of a term relative to the total number of terms in the document.

3.5 Topic Modeling

A topic model is a statistical model that uncovers hidden thematic structures in an extensive collection of documents. It aims to represent each document as a mixture of latent topics [9, 22]. Topic Models produce two byproducts: topic-word matrix and document-topic distribution matrix. Both matrices can be used as features for document representation to train the classifier.

In this context, various topic modeling algorithms can be applied for text classification purposes, including Latent Dirichlet Allocation (LDA) [23], and Latent Semantic Indexing (LSI) [24]. However, given the nature of dealing with app reviews as short texts in this study, we opted to utilize the GSDMM (Gibbs Sampling Dirichlet Mixture Model), known as DMM [25] and BTM (Biterm Topic Model) [26] short text topic models for topic extraction. These models are widely employed in natural language processing and text mining to reveal latent topics within collections of concise texts.

3.6 Leveraging Topic-Modeling Based Machine Learning Algorithms

This study performed text classification through TextNetTopics and TextNetTopics Pro algorithms. TextNetTopics, introduced by Yousef and Voskergian in [10], stands out as an innovative topic modeling-based text classification approach. In contrast to conventional techniques that focus on selecting individual words, TextNetTopics specializes in topic selection. A "topic" is defined as a set of semantically related words derived from a topic model. TextNetTopics selects the top r topics to train the classifier.

On the other hand, TextNetTopics Pro [11] is an advanced version, which is a novel framework oriented towards the short-text classification that uses a promising combination of lexical features organized in topics of words and topic distribution extracted by a short text topic model, effectively mitigating the data-sparseness challenge encountered in the classification of short texts.

The foundation of TextNetTopics and TextNetTopics Pro are rooted in the G-S-M (Grouping, Scoring, and Modeling) approach [27], a generic methodology that found its primary application in the realm of biological data analysis.

3.7 Evaluating Performance Measures

We compared previous models using standard metrics such as Accuracy, Precision, Recall, AUC, Specificity and F1 Measure. The aim was to identify the superior classifier algorithm for app review classifications. The experimental results section will emphasize the F1-score to provide comprehensive insights into model performance.

4 Evaluation

4.1 Experimental Design

In order to derive the topic-word matrix and the document-topic matrix from the GSDMM and BTM algorithms, we utilized STTM [28], an open-source Java library designed for Short Text Topic Modeling.

Both algorithms used fixed hyperparameters ($\alpha = 0.1$ and $\beta = 0.01$) and ran Gibbs sampling inference for 1,000 iterations to ensure convergence. The final samples were used to estimate model parameters. The number of topics and words per topic was set to twenty, as TextNetTopics showed optimal performance among different configurations.

In the context of the document-topic distribution matrix employed by TextNetTopics Pro, the constraint of our relatively small dataset (around 1000 records) raised concerns about the potential suboptimal quality of training a topic model directly. To overcome this limitation, we chose to augment our dataset by merging it with the original dataset, originally containing 100,000 app reviews. Following this, we conducted training on a topic model using the expanded dataset. Afterward, we used only the topic distributions corresponding to the specific app reviews utilized in this study.

We utilized the Knime workflows available in [29] to execute all the preprocessing mentioned earlier. Regarding TextNetTopics and TextNetTopics Pro, their KNIME workflow implementations, available in [29], were employed.

To assess the performance of our approach and gauge its statistical significance, we applied a Stratified Monte Carlo Cross-Validation (MCCV) method, repeated ten times. The dataset was divided into two parts in each iteration: ninety percent for training and ten percent for testing. The use of MCCV ensures that every observation in the dataset has an opportunity to appear in both the training and testing sets. Stratified splitting was utilized, maintaining equal proportions of instances in each class.

In addition to the aforementioned procedures, an essential aspect of our approach involved transforming the original multi-class problem into a binary-class problem. In each time, one class is considered positive with the remaining negative. This transformation was executed systematically, designating one class as positive while treating the remaining classes as negative in each iteration. Additionally, to address potential imbalances between classes, we implemented equal sampling, ensuring a fair representation of instances from each class.

4.2 Experimental Results

According to the results in Table 2, incorporating TextNetTopics with the DMM topic model achieves the highest F1-score with 79.9 and 80.2 for classifying the user experience and bug report app reviews, respectively. On the other hand, incorporating TextNet-Topics with the BTM topic model yields the highest performance, 76.0 and 92.6, respectively, when classifying enhancement requests and text-rating app reviews.

Table 2. Performance comparison of TextNetTopics incorporating DMM and BTM topic models across the four classes. The topic-word matrix used is 20 × 20

Review Type	# of Topics	Topic Model	Terms	Accuracy	Specificity	F-measure	AUC	Recall	Precision
User Experience	10	DMM	78.6	77.9	67.9	**79.9**	85.2	87.9	73.6
	10	BTM	102.6	77.4	70.7	78.8	84.6	84.1	74.3
Bug Report	10	DMM	83.7	80.7	81.3	**80.2**	89.3	80.0	81.1
	9	BTM	93.1	79.8	82.7	79.1	86.2	77.0	81.9
Enhancement Request	10	DMM	79.3	73.9	68.6	75.4	81.5	79.0	72.5
	10	BTM	106.2	74.4	68.6	**76.0**	81.5	80.0	72.6
Text Rating	8	DMM	72.3	92.0	97.0	91.5	94.7	87.0	96.7
	8	BTM	85.6	93.0	97.0	**92.6**	96.3	89.0	96.9

Tables 3 and 4 present the findings of applying TextNetTopics Pro with DMM and BTM topic models at different topic extraction configurations on app reviews. Notably, when configuring the number of topics extracted by a topic model to 20, a substantial enhancement in the F1 score is observed, particularly in the classification of bug report and enhancement request app reviews. For example, a gain of 4.8 and 4.1 is achieved when using the DMM, and 11.2 and 2.4 when using the BTM compared to TextNetTopics. No significant improvement is observed in text rating app reviews, and there is even a decrease in performance for user experience app reviews when using DMM.

However, when configuring the number of topics extracted by a topic model to 50, there is an observed increase in the F1-score, specifically by 2.0 and 2.8 for text rating app reviews when using the DMM and BTM compared to TextNetTopics, respectively. Additionally, a modest improvement of 0.5 is noted when employing DMM for classifying user experience app reviews.

Table 5 summarizes the F1-score improvement when applying TextNetTopics Pro to classify the four types of app reviews over TextNetTopics.

Table 3. Performance comparison of TextNetTopics Pro (#1) incorporating DMM and BTM topic models across the four classes. The utilized metrices are topic-word matrix (20 × 20) and document-topic distribution matrix (n × 20)

Review Type	# of Topics	Topic Model	Terms	Accuracy	Specificity	F-measure	AUC	Recall	Precision
User Experience	10	DMM	96.1	77.2	77.6	77.1	85.7	76.9	78.2
	9	BTM	108.3	77.9	76.6	**78.9**	86.4	79.3	78.5
Bug Report	8	DMM	90.5	84.8	83.3	85.0	92.0	86.3	84.2
	6	BTM	82.4	90.3	90.7	**90.3**	94.7	90.0	90.8
Enhancement Request	7	DMM	82.9	78.7	74.7	**79.5**	86.6	82.7	76.7
	8	BTM	103.2	77.2	71.3	78.4	85.7	83.0	74.5
Text Rating	2	DMM	40	92.9	97.6	92.4	95.1	88.0	97.4
	1	BTM	40	92.2	97.6	**92.9**	95.4	88.9	97.2

Table 4. Performance comparison of TextNetTopics Pro (#2) incorporating DMM and BTM topic models across the four classes. The utilized metrices are topic-word matrix (20 × 20) and document-topic distribution matrix (n × 50)

Review Type	# of Topics	Topic Model	Terms	Accuracy	Specificity	F-measure	AUC	Recall	Precision
User Experience	6	DMM	106	80.2	78.6	**80.4**	88.1	81.7	79.7
	5	BTM	105.9	77.8	76.6	78.0	82.0	79.0	77.6
Bug Report	10	DMM	129.8	84.0	83.7	85.1	93.1	84.3	85.9
	2	BTM	70	88.0	87.7	**88.1**	94.1	88.3	88.0
Enhancement Request	8	DMM	116	79.3	77.3	**79.6**	86.8	81.3	78.3
	7	BTM	126.9	77.3	69.0	79.2	86.1	85.7	74.0
Text Rating	4	DMM	90.9	93.9	96.7	93.5	97.4	91.0	96.3
	6	BTM	109.5	95.6	97.1	**95.4**	98.5	94.0	97.1

Table 5. Performance comparison of TextNetTopics (TNT) and TextNetTopics Pro (TNT PRO) incorporating DMM and BTM topic models across the four classes.

Review Type	Topic Model	TNT	TNT PRO#1	F1-score increase	TNT PRO#2	F1-score increase
User Experience	DMM	79.9	77.1	−2.8	80.4	0.5
	BTM	78.8	78.9	0.1	78.0	−0.8
Bug report	DMM	80.2	85.0	4.8	85.1	4.8

(*continued*)

Table 5. (*continued*)

Review Type	Topic Model	TNT	TNT PRO#1	F1-score increase	TNT PRO#2	F1-score increase
Enhancement Request	BTM	79.1	90.3	11.2	88.1	9.0
	DMM	75.4	79.5	4.1	79.6	4.2
	BTM	76.0	78.4	2.4	79.2	3.2
Text Rating	DMM	91.5	92.4	0.9	93.5	2.0
	BTM	92.6	92.9	0.3	95.4	2.8

5 Conclusion

This study delved into Arabic Text Classification applied to mobile app reviews, explicitly focusing on extracting requirement-related information. The investigation revealed a scarcity of research on the application of Arabic TC for the evolution of mobile app requirements, despite the importance of Arabic as the fourth language used on the Internet and the sixth official language reported by the UN, with approximately 422 million Arabic speakers, emphasizing the novelty and significance of this endeavor.

Our investigation into ATC for mobile app reviews using a topic modeling-based approach with TextNetTopics and TextNetTopics Pro yielded promising outcomes. Notably, integrating TextNetTopics with the DMM topic model demonstrated high F1 scores in classifying user experience and bug request app reviews, while incorporating BTM with TextNetTopics revealed high F1 scores in classifying enhancement requests and text rating app reviews. For TextNetTopics Pro, configuring the number of topics to 20 resulted in substantial improvements in F1 scores, particularly for bug reports and enhancement request app reviews. However, nuanced impacts were observed when increasing the number of topics to 50, indicating the need for further optimization.

We hypothesize that our proposed methodology will be vitally important in product evolution and customization and effectively suggest new ideas for Arabic apps in App Stores. The utilization of app reviews text classifiers will enable developers to make informed decisions on which features to include and what feature to implement first.

In future work, we aim to expand the size of the dataset further to enhance the robustness and generalizability of the obtained results. This expansion aims to capture a more comprehensive representation of the diverse spectrum of mobile app reviews across diverse digital platforms. Additionally, there is an intention to fine-grain the classes, allowing for a more nuanced and detailed categorization of app reviews, which could lead to more precise insights and tailored recommendations for developers.

References

1. Pacheco, C., García, I., Reyes, M.: Requirements elicitation techniques: a systematic literature review based on the maturity of the techniques. IET Softw. **12**, 365–378 (2018)
2. Maalej, W., Nayebi, M., Johann, T., Ruhe, G.: Toward data-driven requirements engineering. IEEE Softw. **33**, 48–54 (2016). https://doi.org/10.1109/MS.2015.153

3. Lim, S., Henriksson, A., Zdravkovic, J.: Data-driven requirements elicitation: a systematic literature review. SN Comput. Sci. **2**, 16 (2021)
4. Groen, E.C., et al.: The crowd in requirements engineering: the landscape and challenges. IEEE Softw. **34**, 44–52 (2017)
5. Annual number of mobile app downloads worldwide 2022. https://www.statista.com/statistics/271644/worldwide-free-and-paid-mobile-app-store-downloads/
6. Mcilroy, S., Shang, W., Ali, N., Hassan, A.E.: User reviews of top mobile apps in Apple and Google app stores. Commun. ACM **60**, 62–67 (2017). https://doi.org/10.1145/3141771
7. Pagano, D., Maalej, W.: User feedback in the appstore: an empirical study. In: 2013 21st IEEE International Requirements Engineering Conference (RE), Rio de Janeiro-RJ, Brazil, pp. 125–134. IEEE (2013)
8. Guzman, E., Maalej, W.: How do users like this feature? A fine grained sentiment analysis of app reviews. In: 2014 IEEE 22nd International Requirements Engineering Conference (RE), Karlskrona, Sweden, pp. 153–162. IEEE (2014)
9. Onan, A., Korukoglu, S., Bulut, H.: LDA-based topic modelling in text sentiment classification: an empirical analysis. Int. J. Comput. Linguist. Appl. **7**, 101–119 (2016)
10. Yousef, M., Voskergian, D.: TextNetTopics: text classification based word grouping as topics and topics' scoring. Front. Genet. **13**, 893378 (2022). https://doi.org/10.3389/fgene.2022.893378
11. Voskergian, D., Bakir-Gungor, B., Yousef, M.: TextNetTopics Pro, a topic model-based text classification for short text by integration of semantic and document-topic distribution information. Front. Genet. **14**, 1243874 (2023)
12. Aslam, N., Ramay, W.Y., Xia, K., Sarwar, N.: Convolutional neural network based classification of app reviews. IEEE Access **8**, 185619–185628 (2020). https://doi.org/10.1109/ACCESS.2020.3029634
13. Panichella, S., Di Sorbo, A., Guzman, E., Visaggio, C.A., Canfora, G., Gall, H.C.: How can i improve my app? Classifying user reviews for software maintenance and evolution. In: 2015 IEEE International Conference on Software Maintenance and Evolution (ICSME), Bremen, Germany, pp. 281–290. IEEE (2015)
14. Maalej, W., Kurtanović, Z., Nabil, H., Stanik, C.: On the automatic classification of app reviews. Requir. Eng. **21**, 311–331 (2016). https://doi.org/10.1007/s00766-016-0251-9
15. Ekanata, Y., Budi, I.: Mobile application review classification for the Indonesian language using machine learning approach. In: 2018 4th International Conference on Computer and Technology Applications (ICCTA), Istanbul, pp. 117–121. IEEE (2018)
16. Al Kilani, N., Tailakh, R., Hanani, A.: Automatic classification of apps reviews for requirement engineering: exploring the customers need from healthcare applications. In: 2019 Sixth International Conference on Social Networks Analysis, Management and Security (SNAMS), Granada, Spain, pp. 541–548. IEEE (2019)
17. Saudy, R.E., Nasr, E.S., El-Ghazaly, A.E.D.M., Gheith, M.H.: Use of Arabic sentiment analysis for mobile applications' requirements evolution: trends and challenges. In: Hassanien, A., Shaalan, K., Gaber, T., Tolba, M. (eds.) AISI 2017. AISC, vol. 639, pp. 477–487. Springer, Cham (2018). https://doi.org/10.1007/978-3-319-64861-3_45
18. Voskergian, D., Jayousi, R.: Identifying anti-vaccination tweets in Arabic language utilizing NLP, ML, and DL approaches. In: 2021 International Conference on Promising Electronic Technologies (ICPET), Deir El-Balah, Palestine, pp. 7–13. IEEE (2021)
19. Voskergian, D., Saheb, M.H.: AMAR_ABSA: Arabic mobile app reviews dataset targeting aspect-based sentiment analysis tasks. In: 2022 Innovations in Intelligent Systems and Applications Conference (ASYU), Antalya, Turkey, pp. 1–7. IEEE (2022)
20. Kanan, T., et al.: A review of natural language processing and machine learning tools used to analyze Arabic social media. In: 2019 IEEE Jordan International Joint Conference on Electrical Engineering and Information Technology (JEEIT), pp. 622–628. IEEE (2019)

21. Eklund, M.: Comparing feature extraction methods and effects of pre-processing methods for multi-label classification of textual data (2018)
22. Barde, B.V., Bainwad, A.M.: An overview of topic modeling methods and tools. In: 2017 International Conference on Intelligent Computing and Control Systems (ICICCS), Madurai, pp. 745–750. IEEE (2017)
23. Blei, D.M., Ng, A.Y., Jordan, M.I.: Latent Dirichlet allocation. J. Mach. Learn. Res. **3**, 993–1022 (2003)
24. Dumais, S.T.: Latent semantic analysis. Annu. Rev. Inf. Sci. Technol. **38**, 188–230 (2004)
25. Yin, J., Wang, J.: A Dirichlet multinomial mixture model-based approach for short text clustering. In: Proceedings of the 20th ACM SIGKDD International Conference on Knowledge Discovery and Data Mining, New York, USA, pp. 233–242. ACM (2014)
26. Yan, X., Guo, J., Lan, Y., Cheng, X.: A biterm topic model for short texts. In: Proceedings of the 22nd International Conference on World Wide Web, Rio de Janeiro Brazil, pp. 1445–1456. ACM (2013)
27. Yousef, M., Kumar, A., Bakir-Gungor, B.: Application of biological domain knowledge based feature selection on gene expression data. Entropy **23**, 2 (2020). https://doi.org/10.3390/e23010002
28. Qiang, J., Qian, Z., Li, Y., Yuan, Y., Wu, X.: Short text topic modeling techniques, applications, and performance: a survey. IEEE Trans. Knowl. Data Eng. **34**, 1427–1445 (2022). https://doi.org/10.1109/TKDE.2020.2992485
29. Yousef, M.: TextNetTopics-Pro (2023). https://github.com/malikyousef/TextNetTopics-Pro

Feature Selection for High-Dimensional Gene Expression Data: A Review

Sara Baali[1]([✉]) [iD], Mohammed Hamim[1] [iD], Hicham Moutachaouik[1], Mustapha Hain[1], and Ismail EL Moudden[2] [iD]

[1] AICSE Laboratory, University Hassan II, Casablanca, Morocco
baalisara92@gmail.com
[2] Eastern Virginia Medical School, EVMS-Sentara Healthcare Analytics and Delivery Science Institute, Norfolk, USA

Abstract. In recent decades, dealing with high-dimensional data has become an undeniable challenge in most data mining applications. In certain domains, such as bio-informatics, and specifically in microarray data analysis, exploring gene expression data often involves the use of tens of thousands of features (genes) measured across just a few dozen samples. Such scenarios, make the use of classical data mining tools a real challenge due to the involvement of a significant number of irrelevant or redundant genes. In response to this challenge, several approaches-based feature selection have been proposed, each with its advantages and disadvantages. This work introduces a classification of feature selection methods and also reviews the state-of-the-art approaches developed over the past five years. Our review has revealed a notable trend towards hybrid approaches, approximately 50% of the surveyed studies propose hybrid feature selection techniques, most frequently combining filter with wrapper methods. Additionally, the 10-fold cross-validation technique stands out as the dominant evaluation method, employed by 61.6% of surveyed approaches. Support Vector Machines emerge as the most favored classification algorithm, demonstrating optimal performance in 77.78% of cases. These findings contribute to the advancement of feature selection approaches, particularly in reducing the dimensionality of gene expression data, thereby enhancing cancer classification methodologies.

Keywords: Gene Expression Data · Feature Selection · Classification

1 Introduction

While many Machine Learning (ML) techniques are adapted to handle a large amount of data, their performance degrades as the dimensionality of the data increases. Indeed, as the number of variables rises, the number of observations increases proportionally, and the dimensionality becomes more significant. Consequently, data visualization and comprehension become much more complex, computation time and resource consumption increase, and learning models inadequately describe the studied problem due to the involvement of irrelevant or redundant features [1, 2]. This phenomenon, known as the curse of dimensionality, a term first introduced by Richard E. Bellman [3], is a challenge present in almost all data mining applications. Cancer diagnosis using gene expression data is an application that frequently grapples with this challenge. The analysis of gene expression data for cancer identification often involves dealing with thousands of irrelevant or redundant genes, necessitating the implementation of an appropriate and efficient approach for their removal. Feature Selection (FS) can address this issue by selecting the most relevant genes, thereby enhancing classification performance and reducing the complexity of cancer identification models, which in turn, makes the models more interpretable for domain experts [2].

The main objective of this study is to present a systematic review by examining a set of FS approaches, with a particular focus on reducing the dimensionality of gene expression data over the last five years.

The structure of the remaining sections in this paper is outlined as follows: the next section provides an overview of different types of feature selection techniques. Section 3 provides a review of microarray feature selection methods developed between 2019 and 2023. Section 4 provides a quantitative examination of the literature's feature selection approaches, while the conclusion and perspectives are discussed in the final section.

2 Background

2.1 DNA Microarray Technology and Gene Expression Data

The adoption of Deoxyribonucleic Acid (DNA) microarray technology had a significant impact on molecular biology in recent decades. Developed in the early 1990s, DNA microarray technology allowed the scientific community, for the first time, to obtain a comprehensive view of the cell and study its molecular biological state by simultaneously measuring the expression of tens of thousands of genes [4, 5]. This powerful technology involves utilizing solid surfaces, typically glass slides or silicon chips, onto which thousands of DNA probes are attached, representing specific genomic sequences. Consequently, the gene expression data generated by microarray technology is organized in a table format, where the rows correspond to individual genes, and the columns signify instances that represent different patients.

2.2 Challenges of Gene Expression Data and Feature Selection Methods

Analyzing gene expression data for cancer classification poses several challenges in the field of molecular biology. Indeed, the presence of a limited number of samples, often on the order of a few dozen, against a huge number of genes/features, typically in the thousands, introduces the risk of overfitting, making it challenging to develop robust and generalizable cancer classification models [6]. In this context, feature/gene selection plays a crucial role in mitigating overfitting risks, simplifying the interpretability of classification models, and improving the overall accuracy and reliability of cancer classification [7]. In the context of classification, FS methods are mainly organized into four categories: filter, wrapper, embedded, and hybrid methods.

Filter: Instead of involving an ML approach to evaluate feature relevance, these methods (see Fig. 1) use feature ranking characteristics as their main selection criteria [8]. A filter method measures feature relevance utilizing various types of evaluation criteria such as correlation, dependency, degree of class separability, or distance and consistency [9].

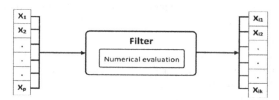

Fig. 1. *Filter* method for feature selection [9]

These methods are applied before the learning process and aim to identify and select features that are likely to be informative for the target variable, the reason why these methods are generally considered much faster than other selection approaches, and more suitable for handling extensive datasets. However, their major drawback is their failure to consider the impact of the selected subsets on learning performance [10].

Wrapper: Introduced by Kohavi and John, these methods (see Fig. 2) are designed to address the limitation of filter methods, which often overlook the influence of the selected features on the classifier's performance [11]. It wraps the FS process around the model evaluation, treating it as an optimization problem. The key idea is to evaluate different subsets of features by training and testing the model using a specific learning algorithm. The performance of the model with each subset is used as a criterion for selecting the best set of features.

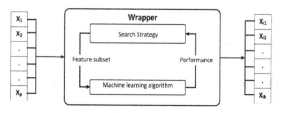

Fig. 2. *Wrapper* method for feature selection [9]

With this strategy, the wrapper method often shows better performance compared to the filter method, as demonstrated by Kohavi and John [11]. However, among other things, two main drawbacks prevent the wrapper method from being the perfect selection approach. The first issue is the complexity of this technique, which increases exponentially when dealing with a large number of features. This makes the use of an exhaustive search strategy impractical. The second issue is that the quality of the selected subset is specific to the ML algorithm used in the evaluation procedure. In other words, if we evaluate the same selected subset with another classifier, we may not achieve the same performance. Sequential selection algorithms and metaheuristic algorithms are among the two main types of wrapper methods.

Embedded: In these methods, the FS process is completely integrated as part of the model training process. Unlike filter and wrapper methods, embedded methods select features during the training phase of the ML model. This allows them to consider not only potential relationships between features and the target class but also to conduct local searches for features that enhance classification or regression performance [12]. However, these methods, similar to other selection techniques, also have their weaknesses, one of which is that the quality of the selected subset is specific to the learning process used.

Hybrid: These methods involve integrating aspects of various feature selection methods, such as filter, wrapper, or embedded methods to capitalize on their individual strengths and address their respective limitations. The main aim is to establish a more robust and effective FS process. The most frequently adopted combination involves filters and wrappers. In this approach, the simplicity and speed of the filter method are utilized to reduce the size of the feature space. Subsequently, the wrapper is employed to identify subsets that exhibit the best classification performance [13].

3 A Review on Microarray Feature Selection

Recently, various methods for reducing dimensionality have been developed to tackle the issue of high dimensionality in DNA microarray data. Each approach has its own set of strengths and weaknesses. This section provides an overview of recent studies explored in our literature review:

In 2019: Alanni et al. introduced GSP (Gene Selection Programming), an innovative gene selection method employing a Support Vector Machine (SVM) with a linear kernel as the classifier. The method underwent testing on ten gene expression datasets using Leave-One-Out Cross-Validation (LOOCV), with accuracy as the sole metric for classification performance [14]. Yuan et al. introduced a novel ranking-based FS method, PMCI (Partial Maximum Correlation Information). The proposed method involves extracting multiple orthogonal components from the feature space by considering the correlation between the feature space and the class representation space [15]. In their work, Kang et al. introduced an innovative tumor classification methodology, termed rL-GenSVM, representing the combination of relaxed Lasso (Least Absolute Shrinkage and Selection Operator) with the Generalized Support Vector Machine. The initial step involves z-score normalization of the tumor dataset. Then, a relaxed Lasso procedure is implemented to select subset of the relevant gene from the training set. Finally, GenSVM is employed as the classifier [16].

In 2020: Shukla introduced EMPAGA (Ensemble MultiPopulation Adaptive Genetic Algorithm) for gene selection, employing a two-phase approach. The first phase, EGS (Ensemble Gene Selection), employs a novel filter combining statistical measures and Information Gain to reduce research space. The second phase utilizes an adaptive genetic algorithm with SVM and Naive Bayes (NB) as fitness functions, demonstrating competitive dimensionality reduction and enhanced cancer classification performance [17]. Wahid et al. have introduced a new filter-based approach named Correlation-based Overlapping Score (COS). The proposed approach employs the Symmetrical Uncertainty (SU) measure to identify pertinent features that predominantly influence the target class. The authors exclusively tested their algorithm on datasets with binary class [18].

In 2021: Babu et al. proposed a hybrid gene selection method integrating an unsupervised filter and a wrapper with a heuristic search technique. The filter involves clustering genes using k-means, ranking them based on Signal-to-Noise Ratio (SNR), and selecting the top genes from each cluster. The wrapper refines the selection using Cellular Learning and Ant Colony Optimization (CLACO). The method's performance was assessed with SVM, k-nearest neighbors (K-NN), and NB classifiers [19]. In the same context, Kafrawy et al. introduced SVM-mRMRe, a hybrid method that combines mRMR (minimum Redundancy–Maximum Relevancy)-based filter method with SVM-based embedded method. SVM initially reduced the gene space, followed by mRMRe filtering for redundancy. Merging the outcomes, a voting system identified the most relevant genes [20]. In the same year, Houssein et al. introduced BMO (Barnacles Mating Optimizer), a metaheuristic optimization algorithm for feature selection using SVM in the fitness function. The approach showed high performance in accuracy and gene selection, with emphasis on testing it with the same classifier in the fitness function [21].

In 2022: Chamlal et al. proposed a hybrid filter-wrapper preordinances-based FS named Preordonnances and Complementarity in terms of new Relevance in Weighted Graphs (PCRWG). In the filter step, they introduce a scoring system to quantify both Relevance and Complementarity. While during the wrapper step, an undirected Weighted Graph is constructed using a relevance-complementarity matrix [22]. In a similar vein, Chaudhuri and Sahu developed another hybrid approach, QOMOJaya (Quasi-Oppositional based Multi-Objective Jaya algorithm). In its filter step, only 2% of genes were retained through five filters, and then subjected to the QOMOJaya algorithm for selecting the optimal gene subset [23]. Kundu et al. developed a Meta-heuristic approach based on the Altruistic Whale Optimization Algorithm (AltWOA) [24]. Rostami et al. introduced an innovative filter-based gene selection method by incorporating the concept of community detection with node centrality (CDNC). The proposed approach integrates relevance and similarity criteria into its selection search strategy [25]. In the same context of Meta-heuristic-based approaches, Adebayo et al. presented an innovative hybrid method for gene selection in microarray data, fusing a data-driven threshold algorithm with the optimization capabilities of Fuzzy Particle Swarm Optimization (FPSO) [26].

In 2023, Akhavan et al. introduced a two-phase gene selection approach. Initially, they reduced the gene space using an anomaly (outliers) detection-based method. In the second phase, a guided genetic algorithm refines and identifies a subset of the most relevant genes [27]. Alhenawi et al. developed a hybrid approach that combines an ensemble filter with an improved Intelligent Water Drop (IWD) algorithm, incorporating one of three local search (LS) algorithms (Tabu search, Novel LS algorithm, Hill Climbing) in each IWD iteration. Their proposed approach implements the correlation coefficient as a heuristic undesirability (HUD) for selecting the next node in the original IWD algorithm

[28]. In their paper, Xie et al. proposed an enhanced multilayer binary firefly-based hybrid method. Initially, they reduced the research space using an improved mRMR-based filter method. Then, the reduced space was further refined from a coarse-grained to a fine-grained level through the improved multilayer binary firefly algorithm (MBFA) [29]. Lee et al. proposed a novel multivariate feature ranking method called Markov Blanket (MB). Their main idea is to incorporate the formal definition of relevance into the MB, transforming it into a multivariate ranking method. This approach simultaneously considers relevance to the target and redundancy among features within reasonable time and memory complexity [30]. Pati et al. introduced a hybrid approach named Heatmap Analysis and Graph Neural Network (HAGNN). In their approach the FS starts with a heatmap analysis applied to various classes of microarray data, resulting in the identification of Regions of Interest (ROIs). Subsequently, the Graph Neural Network (GNN) is used as an embedded method that automatically learns and extracts pertinent features from the data [31].

Table 1 illustrates a comprehensive overview of gene selection approaches categorized by FS type (wrapper, embedded, filter, and hybrid). The Method column denotes the name of each proposed approach used for gene selection. The Datasets column outlines the gene expression datasets suggested for assessing each gene selection approach, providing insight into the diverse data sources considered. The #Selected Features column represents the average number of selected features. Classifier enumerates the classifiers used to evaluate the approach's ability to classify cancer, and Accuracy represents the achieved classification performance. Lastly, the Evaluation Techniques column provides a view of the assessment strategies applied in each research paper.

4 Analysis of Reviewed Feature Selection Approaches

4.1 Approaches Distribution Analysis

This section represents a distribution of approaches analysis in terms of the feature selection type, the classifiers that have generated the best results, and the evaluation techniques. Figure 3 illustrates the distribution of the approaches in terms of different feature selection methods. As depicted, it is evident that the majority of works focus on proposing hybrid selection approaches, representing 50% of the reviewed approaches, attempting to combine multiple selection methods rather than relying on singular approaches. The common combination involves filter and wrapper methods. In contrast, embedded methods are the least adopted, comprising only 5.56% of the reviewed approaches.

Figure 4 illustrates the distribution of reviewed FS approaches based on the adopted evaluation technique. As observed, the majority of reviewed approaches tend to adopt the 10-fold CV technique, representing approximately 77.78% of approaches, while the 5-fold CV technique is less commonly adopted, representing only 5.56% of all reviewed papers.

Table 1. Comparison of feature selection methods used in gene expression data classification from 2019 to 2023.

Ref.	FS	Method	Datasets	#S	#G	#C	#SF	Cls	A (%)	Ev
[14]	wrapper	GSP	11_Tumors	174	12533	11	17.9	SVM	99.88	LOOCV
			9_Tumors	60	5726	11	13.8		98.88	
			Brain Tumor1	90	5920	5	9.2		99.8	
			Brain Tumor2	50	10367	4	9.8		99.9	
			Leukemia1	72	5327	3	2.9		100	
			Leukimia2	72	11225	3	4.1		100	
			Lung	203	12600	5	8.3		100	
			SRBCT	82	2308	4	4		100	
			Prostate	102	10509	2	8.2		99.87	
			DLBCL	77	5469	2	3.5		100	
[15]	Filter	PMCI	GLI-85	85	22283	2	14	SVM	92.94	LOOCV
			Prostate-GE	102	5966	2	8	RF	95.10	
			Leukemia	72	7129	2	60	RF	98.61	
			CLL-SUB	111	11340	3	86	RF	87.39	
			Breast	95	4869	3	58	SVM	74.74	
			Lymphoma	62	4026	3	20	SVM	100	
			GLIOMA	50	4434	4	49	RF	84	
			Brain	42	5597	5	34	RF	90.48	
			Lung	203	12600	5	28	RF	96.06	
			GCM	198	16063	14	96	RF	75.76	

(*continued*)

Table 1. (*continued*)

Ref.	FS	Method	Datasets	#S	#G	#C	#SF	Cls	A (%)	Ev
[16]	Embedded	rL-GenSVM	DLBCL	77	7129	2	9	GenSVM	93.33	10-flod cross-Validation(CV)
			CNS	60	7129	2	16		75.00	
			Lung	86	7129	2	21		76.47	
			Ovarian	253	15155	2	3		100	
			Brain	50	10367	4	8		100	
			Lymphoma	62	4026	3	14		92.86	
			MLL	72	12582	3	28		60.00	
			TOX-171	171	5748	4	29		76.47	
[17]	Hybrid	EMPAGA	Breast	97	24481	2	16.7	SVM	95.63	10-flod CV
			Colon	62	2000	2	15	NB	97.23	
			DLBCL	77	5469	2	18.1		99.23	
			SRBCT	83	2308	4	12.1		98.99	
			Leukemia	72	7129	2	13.1		98.92	
			Lung	203	12600	5	13.2		99.63	
			Brain Tumors1	90	5920	5	16.27		95.62	
			11_Tumors	174	12533	11	12.5		91.23	
			9_Tumors	60	5726	9	16.25		83.63	
			Prostate	102	10509	2	15.52		97.23	
[18]	Filter	COS	Leukemia	72	7129	2	2	Boosting	100	10-flod CV
			Lung	181	12533	2	3	KNN	98.60	
			Prostate	102	12600	2	–	–	–	
			Colon	62	2000	2	–	–	–	
			DLBCL	77	5469	2	35	Boosting	96.00	

(*continued*)

Table 1. (*continued*)

Ref.	FS	Method	Datasets	#S	#G	#C	#SF	Cls	A (%)	Ev
[19]	Hybrid	KM-SNR-CLACO	Leukemia	72	7129	2	14.6	NB	100	10-flod CV
			Prostate	136	12600	2	56	KNN	100	
			DLBCL	77	5469	2	26.3	KNN	99.42	
			MLL	72	12582	3	27.6	NB	100	
			SRBCT	83	2308	4	46.4	KNN	100	
[20]	Filter	SVM-mRMRe	Prostate	102	12600	2	65	SVM	99.03	8-fold CV
			Lung	181	12533	2	292	KNN	100	
			Brain	28	1070	2	16	RF	100	
			Colon	62	2000	2	40	Multi-Layer Perceptron (MLP)	98.05	
			CNS	60	7129	2	33		100	
			Breast	49	7129	2	64		100	
			Leukemia	72	7129	2	40		100	
			Ovarian	253	15155	2	6		100	
[21]	Filter	BMO-SVM	Leukemia	72	7129	2	3	SVM	100	10-fold CV
			SRBCT	83	2308	4	4		100	
			Lymphoma	62	4026	3	2		100	
			Leukemia2	72	7129	3	N/A		100	
[22]	Hybrid	PCRWG	Colon	62	2000	2	5	SVM	90.53	10-fold CV
			Leukemia	72	3571	2	9		100.00	
			Breast	78	4348	2	2		81.08	

(*continued*)

Table 1. (continued)

Ref.	FS	Method	Datasets	#S	#G	#C	#SF	Cls	A (%)	Ev
			CNS	60	7129	2	2		78.13	
			ALL_AML_2C	72	7129	2	6		100.00	
			Lung	181	12534	2	5		100.00	
			Prostate	102	12600	2	23		99.14	
			Ovarian	253	15154	2	36		100.00	
			Lymphoma	62	4026	3	2		100.00	
			ALL_AML_3C	72	7129	3	2		97.50	
			MLL	72	12582	3	27		97.08	
			SRBCT	83	2309	4	30		100.00	
			Brain	42	5597	5	20		91.43	
[23]	Hybrid	QOMOJaya	Colon	62	2000	2	13.6	SVM	99.97	10-fold CV
			CNS	60	7129	2	60.4		99.99	
			ALL-AML	72	7129	2	56.8		99.99	
			ALL-AML-3C	72	7129	3	57.9		100	
			ALL-AML-4C	72	7129	4	59.6		100	
			Lung	181	12533	2	117.9		99.93	
			Lymphoma	62	4026	3	11.9		99.98	
			MLL	72	12582	3	109.1		100	
			Ovarian	253	15154	2	131.3		99.99	
[24]	wrapper	AltWOA	Alon	62	2000	2	21	SVM	100	5-fold CV

(continued)

Table 1. (*continued*)

Ref.	FS	Method	Datasets	#S	#G	#C	#SF	Cls	A (%)	Ev
			Christensen	217	1413	3	2		100	
			Gravier	168	2905	2	44		94.12	
			Khan	63	2308	4	25		100	
			Sorlie	85	456	5	35		100	
			Leukemia	72	7129	2	30		100	
			11_Tumors	174	12534	11	37		100	
			DLBCL	77	5470	2	22		100	
[25]	Filter	CDNC	Colon	62	2000	2	13.45	SVM	88.89	hold-out
			SRBCT	83	2309	4	17.43		82.79	
			Leukemia	72	7129	2	21.98		91.16	
			Prostate	102	10509	2	22.81		83.91	
			Lung	181	12533	2	28.21		91.82	
[26]	Hybrid	FPSO	Leukemia	72	7129	2	19	NB	95.83	10-fold CV
			Prostate	102	12600	2	13	KNN & SVM	95.10	
			Lung	181	12533	2	14	NB	99.44	
			Colon	62	2000	2	8	NB	87.10	
[27]	Hybrid	Anomaly Detection (AD) + guided GA	Gastric	30	4522	2	≈1%	SVM	100	10-fold CV
			Leukemia	72	7129	2	''		98.6	
			DLBCL	77	7129	2	''		93.2	
			Prostate	102	12533	2	''		92.8	

(*continued*)

Table 1. (*continued*)

Ref.	FS	Method	Datasets	#S	#G	#C	#SF	Cls	A (%)	Ev
			Ovarian	253	15154	2	"		99.9	
			MLL	72	12582	3	"		97.7	
			SRBCT	83	2309	4	"		97.4	
			Lung	203	12600	5	"		94.2	
[28]	Hybrid	PHFS-IWD	Colon	62	2000	2	2	NB	100	10-fold CV
			DLBCL	47	4026	2	2		100	
			CNS	60	7129	2	5		93.3	
			Prostate	102	12600	2	2		96.15	
			Ovarian	253	15154	2	2		100	
[29]	Hybrid	mRMR-MBFA	Colon	62	2000	2	9.4	–	90.48	10-fold CV
			DLBCL	77	7129	2	5.9	–	98.75	
			Leukemia	72	7129	2	4.3	–	100	
			Prostate	102	12600	2	6.7	–	94.18	
[30]	Filter	Inter-IAMBR	Leukemia	72	7129	2	10	SVM	97.22	LOOCV
			Prostate	102	12600	2	10	SVM	99.02	
			SRBCT	83	2309	4	10	SVM & KNN	100	
			Lung	203	12600	5	10	SVM	94.76	
			Ovarian	253	15154	2	10	SVM & KNN	100	
			MLL	72	12582	3	10	SVM	97.02	

(*continued*)

Table 1. (*continued*)

Ref.	FS	Method	Datasets	#S	#G	#C	#SF	Cls	A (%)	Ev
[31]	Hybrid	HAGNN	Leukemia	72	7129	2	73	SVM	99.561	Hold-out
			Colon	62	2000	2	28	SVM	98.499	
			DLBCL	77	7129	2	92	SVM	99.052	
			Lung	181	12533	2	89	KNN	99.883	
			Prostate	136	12600	2	95	KNN	98.854	

Ref.: paper reference.
#S: number of samples.
#G: number of genes.
#C: number of classes.
#SF: Number of selected features.
Cls: classifier.
A: Accuracy.
Ev: Evaluation technique.

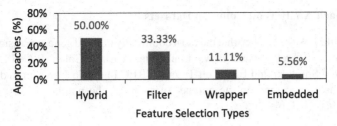

Fig. 3. Distribution of feature selection methods in state-of-the-art

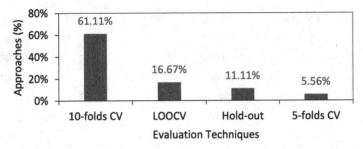

Fig. 4. Distribution of evaluation techniques usage in state-of-the-art approaches.

Figure 5 illustrates the distribution of reviewed FS approaches based on the employed classifiers, emphasizing that these classifiers are among the best-performing ones across all reviewed approaches. It is evident that the SVM emerges as the most effective classifier, with a prevalence of 77.78% across all reviewed approaches, followed closely by KNN, representing 33.33% of the reviewed works. In contrast, MLP appears to be the least employed classifier, with a percentage not exceeding 6% across all reviewed approaches.

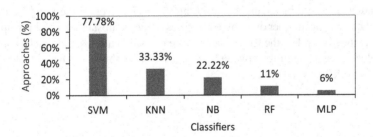

Fig. 5. Distribution of classifiers usage in the state-of-the-art approaches.

4.2 Approach Analysis on Common Datasets

Based on Table 1, we can notice that the most commonly utilized gene expression datasets in all reviewed research are as follows: Leukemia, Prostate, Colon, B-Cell Lymphoma (DLBCL), and Small Round Blue Cell Tumor (SRBCT). Table 1 presents a description of these datasets, focusing on the number of samples, the number of genes, and the number of classes (Table 2).

Table 2. Key characteristics of commonly used gene expression dataset.

Dataset	#Simples	#Gene	#Classes	Ref	Year
Leukemia	72	7129	2	[32]	1999
Colon	62	2000	2	[33]	1999
Prostate	102	12600	2	[34]	2002
DLBCL	77	5469	2	[35]	2000
SRBCT	83	2309	4	[36]	2001
Lung	181	12533	2	[37]	2002

Figure 6 presents a comparison of experimental results from reviewed approaches in terms of the highest achieved accuracies and the corresponding number of involved genes for the most commonly used gene expression dataset described in Table 1. As it can be observed, for the leukemia dataset, we can notice that the most of the reviewed approaches using this data achieved an excellent classification performance by involving a limited number of features that do not exceed 46. The most striking results concern the BMO-SVM and mRMR-MBFA approaches, achieving an accuracy of 100% using only 3 and 4 selected features, respectively. Almost identical observations remain valid for the SRBCT, Lung Cancer, DLBCL, and Prostate datasets, where approximately 70% of the studies achieved an accuracy exceeding 95%. However, in the Lung Cancer and Prostate datasets, the number of involved features remains substantial, reaching up to 290 and 120, respectively. For the Colon dataset, we can say that most approaches achieved acceptable accuracy, ranging between 90% and 100%, with limited features that do not exceed 40, even the worst-case scenarios.

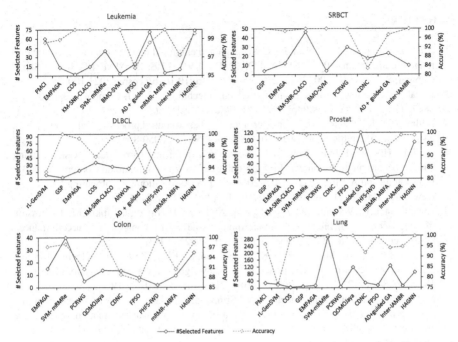

Fig. 6. Highest accuracy and the number of selected genes for each dataset across all reviewed feature selection approaches.

5 Conclusion

This review explored feature selection approaches developed to address the challenges of high-dimensionality in gene expression microarray datasets from 2019 to 2023, with a focus on those published in Elsevier and IEEE. In our review, we aimed to diversify the investigated approaches by considering different types of feature selection methods, various machine learning algorithms, diverse evaluation techniques, as well as the datasets used to assess the effectiveness of the proposed approaches. Our review has uncovered a discernible trend towards hybrid approaches, approximately 50% of surveyed papers, often combine filter with wrapper methods. In contrast, embedded methods were found to be the least used, representing only 5.56% of all investigated papers. Furthermore, the 10-fold CV technique emerged as the predominant evaluation method, employed by 61.6% of surveyed approaches. When it comes to the suggested classifiers, SVM stood out as the most favored one, showcasing optimal performance in 77.78% of cases. These findings contribute to a nearly comprehensive understanding of the landscape of feature selection in high-dimensional gene expression microarray datasets.

Although many publications on feature selection for gene expression data have been examined in the literature, many other approaches need to be investigated to gain a more comprehensive insight. In our future work, we aim to expand the spectrum of explored approaches to validate our conclusions and propose optimal strategies for refining the reduction of high-dimensionality in gene expression data, ultimately leading to enhanced cancer classification.

References

1. Anowar, F., Sadaoui, S., Selim, B.: Conceptual and empirical comparison of dimensionality reduction algorithms (PCA, KPCA, LDA, MDS, SVD, LLE, ISOMAP, LE, ICA, t-SNE). Comput. Sci. Rev. **40**, 100378 (2021). https://doi.org/10.1016/j.cosrev.2021.100378
2. Jindal, P.: A review on dimensionality reduction techniques. Int. J. Comput. Appl. (0975 – 8887) **173**(2) (2017). https://doi.org/10.5120/ijca2017915260
3. Bellman, R., Kalaba, R.: A note on interrupted stochastic control processes. Inf. Control **4**(4), 346–349 (1961). https://doi.org/10.1016/S0019-9958(61)80050-8
4. Rafii, F., Hassani, B.D.R., Kbir, M.A.: New approach for microarray data decision making with respect to multiple sources. In: Proceedings of the 2nd International Conference on Big Data, Cloud and Applications, BDCA 2017, pp. 1–5. Association for Computing Machinery, New York, March 2017. https://doi.org/10.1145/3090354.3090463
5. Augenlicht, L.H., Wahrman, M.Z., Halsey, H., Anderson, L., Taylor, J., Lipkin, M.: Expression of cloned sequences in biopsies of human colonic tissue and in colonic carcinoma cells induced to differentiate in vitro. Cancer Res. **47**(22), 6017–6021 (1987)
6. Hamraz, M., Ali, A., Mashwani, W.K., Aldahmani, S., Khan, Z.: Feature selection for high dimensional microarray gene expression data via weighted signal to noise ratio. PLoS ONE **18**(4), e0284619 (2023). https://doi.org/10.1371/journal.pone.0284619
7. Mahendran, N., Durai Raj Vincent, P.M., Srinivasan, K., Chang, C.-Y.: Machine learning based computational gene selection models: a survey, performance evaluation, open issues, and future research directions. Front. Genet. **11**, 603808 (2020). https://doi.org/10.3389/fgene.2020.603808
8. Guyon, I., Elisseeff, A.: An introduction to feature extraction. In: Guyon, I., Nikravesh, M., Gunn, S., Zadeh, L.A. (eds.) Feature Extraction. SFSC, vol. 207, pp. 1–25. Springer, Heidelberg (2006). https://doi.org/10.1007/978-3-540-35488-8_1
9. Ang, J., Mirzal, A., Haron, H., Hamed, H.N.A.: Supervised, unsupervised, and semi-supervised feature selection: a review on gene selection. IEEE/ACM Trans. Comput. Biol. Bioinform. **13**, 1 (2015). https://doi.org/10.1109/TCBB.2015.2478454
10. Lazar, C., et al.: A survey on filter techniques for feature selection in gene expression microarray analysis. IEEE/ACM Trans. Comput. Biol. Bioinform. **9**(4), 1106–1119 (2012). https://doi.org/10.1109/TCBB.2012.33
11. Kohavi, R., John, G.H.: Wrappers for feature subset selection. Artif. Intell. **97**(1–2), 273–324 (1997). https://doi.org/10.1016/S0004-3702(97)00043-X
12. Aziz, R., Verma, C.K., Srivastava, N., Aziz, R., Verma, C.K., Srivastava, N.: Dimension reduction methods for microarray data: a review. AIMS Bioeng. **4**(1), 179–197 (2017). https://doi.org/10.3934/bioeng.2017.1.179
13. Rothe, F., et al.: Fluorescence optical imaging feature selection with machine learning for differential diagnosis of selected rheumatic diseases. Front. Med. **10**, 1228833 (2023). https://doi.org/10.3389/fmed.2023.1228833
14. Alanni, R., Hou, J., Azzawi, H., Xiang, Y.: A novel gene selection algorithm for cancer classification using microarray datasets. BMC Med. Genomics **12**(1), 10 (2019). https://doi.org/10.1186/s12920-018-0447-6
15. Yuan, M., Yang, Z., Ji, G.: Partial maximum correlation information: a new feature selection method for microarray data classification. Neurocomputing **323**, 231–243 (2019). https://doi.org/10.1016/j.neucom.2018.09.084
16. Kang, C., Huo, Y., Xin, L., Tian, B., Yu, B.: Feature selection and tumor classification for microarray data using relaxed Lasso and generalized multi-class support vector machine. J. Theor. Biol. **463**, 77–91 (2019). https://doi.org/10.1016/j.jtbi.2018.12.010

17. Shukla, A.K., Singh, P., Vardhan, M.: Gene selection for cancer types classification using novel hybrid metaheuristics approach. Swarm Evol. Comput. **54**, 100661 (2020). https://doi.org/10.1016/j.swevo.2020.100661
18. Wahid, A., et al.: Feature selection and classification for gene expression data using novel correlation based overlapping score method via Chou's 5-steps rule. Chemom. Intell. Lab. Syst. **199**, 103958 (2020). https://doi.org/10.1016/j.chemolab.2020.103958
19. Babu P, S.A., , Annavarapu, C.S.R., Dara, S.: Clustering-based hybrid feature selection approach for high dimensional microarray data. Chemom. Intell. Lab. Syst. **213**, 104305 (2021). https://doi.org/10.1016/j.chemolab.2021.104305
20. Kafrawy, P.E., Fathi, H., Qaraad, M., Kelany, A.K., Chen, X.: An efficient SVM-based feature selection model for cancer classification using high-dimensional microarray data. IEEE Access **9**, 155353–155369 (2021). https://doi.org/10.1109/ACCESS.2021.3123090
21. Houssein, E.H., Abdelminaam, D.S., Hassan, H.N., Al-Sayed, M.M., Nabil, E.: A hybrid barnacles mating optimizer algorithm with support vector machines for gene selection of microarray cancer classification. IEEE Access **9**, 64895–64905 (2021). https://doi.org/10.1109/ACCESS.2021.3075942
22. Chamlal, H., Ouaderhman, T., Rebbah, F.E.: A hybrid feature selection approach for microarray datasets using graph theoretic-based method. Inf. Sci. **615**, 449–474 (2022). https://doi.org/10.1016/j.ins.2022.10.001
23. Chaudhuri, A., Sahu, T.P.: Multi-objective feature selection based on quasi-oppositional based Jaya algorithm for microarray data. Knowl. Based Syst. **236**, 107804 (2022). https://doi.org/10.1016/j.knosys.2021.107804
24. Kundu, R., Chattopadhyay, S., Cuevas, E., Sarkar, R.: AltWOA: altruistic whale optimization algorithm for feature selection on microarray datasets. Comput. Biol. Med. **144**, 105349 (2022). https://doi.org/10.1016/j.compbiomed.2022.105349
25. Rostami, M., Forouzandeh, S., Berahmand, K., Soltani, M., Shahsavari, M., Oussalah, M.: Gene selection for microarray data classification via multi-objective graph theoretic-based method. Artif. Intell. Med. **123**, 102228 (2022). https://doi.org/10.1016/j.artmed.2021.102228
26. Adebayo, P.O., Jimoh, R.G., Yahya, W.B.: Hybridization of data-driven threshold algorithm with fuzzy particle swarm optimization technique for gene selection in microarray data. Sci. Afr., e02012 (2023). https://doi.org/10.1016/j.sciaf.2023.e02012
27. Akhavan, M., Hasheminejad, S.M.H.: A two-phase gene selection method using anomaly detection and genetic algorithm for microarray data. Knowl. Based Syst. **262**, 110249 (2023). https://doi.org/10.1016/j.knosys.2022.110249
28. Alhenawi, E., Al-Sayyed, R., Hudaib, A., Mirjalili, S.: Improved intelligent water drop-based hybrid feature selection method for microarray data processing. Comput. Biol. Chem. **103**, 107809 (2023). https://doi.org/10.1016/j.compbiolchem.2022.107809
29. Xie, W., Wang, L., Yu, K., Shi, T., Li, W.: Improved multi-layer binary firefly algorithm for optimizing feature selection and classification of microarray data. Biomed. Signal Process. Control **79**, 104080 (2023). https://doi.org/10.1016/j.bspc.2022.104080
30. Lee, J., Choi, I.Y., Jun, C.-H.: An efficient multivariate feature ranking method for gene selection in high-dimensional microarray data. Expert Syst. Appl. **166**, 113971 (2021). https://doi.org/10.1016/j.eswa.2020.113971
31. Pati, S.K., Banerjee, A., Manna, S.: Gene selection of microarray data using heatmap analysis and graph neural network. Appl. Soft Comput. **135**, 110034 (2023). https://doi.org/10.1016/j.asoc.2023.110034
32. Golub, T.R., et al.: Molecular classification of cancer: class discovery and class prediction by gene expression monitoring. Science **286**(5439), 531–537 (1999). https://doi.org/10.1126/science.286.5439.531

33. Alon, U., et al.: Broad patterns of gene expression revealed by clustering analysis of tumor and normal colon tissues probed by oligonucleotide arrays. Proc. Natl. Acad. Sci. U. S. A. **96**(12), 6745–6750 (1999). https://doi.org/10.1073/pnas.96.12.6745
34. Singh, D., et al.: Gene expression correlates of clinical prostate cancer behavior. Cancer Cell **1**(2), 203–209 (2002). https://doi.org/10.1016/s1535-6108(02)00030-2
35. Alizadeh, A.A., et al.: Distinct types of diffuse large B-cell lymphoma identified by gene expression profiling. Nature **403**(6769), 503–511 (2000). https://doi.org/10.1038/35000501
36. Khan, J., et al.: Classification and diagnostic prediction of cancers using gene expression profiling and artificial neural networks. Nat. Med. **7**(6), 673–679 (2001). https://doi.org/10.1038/89044
37. Gordon, G.J., et al.: Translation of microarray data into clinically relevant cancer diagnostic tests using gene expression ratios in lung cancer and mesothelioma. Cancer Res. **62**(17), 4963–4967 (2002)

Legal Contract Quality and Validity Assessment Through the Bayesian Networks

Youssra Amazou[✉], Abdellah Azmani, and Monir Azmani

Intelligent Automation and BioMedGenomies Laboratory (IAMBL), Faculty of Sciences and Technologies of Tangier, Abdelmalek Essaadi University of Tetouan, Tetouan, Morocco
Amazou.youssra@etu.uae.ac.ma

Abstract. This article examines the fundamental elements of a valid contractual relationship, focusing on mutual consent, fair consideration, legal capacity and compliance with relevant laws. Beyond these fundamental components, it highlights the importance of contract quality in terms of drafting precision and clarity. A well-drafted contract is essential to avoid any ambiguities in the language that could undermine its validity. Furthermore, in the midst of legislative change, it is imperative to maintain compliance with applicable laws. To address these challenges, the paper proposes the use of Bayesian networks to model contractual risks, emphasising the need for robust data collection and precise probability specification for each variable. The primary goal of this approach is to assist legal professionals in evaluating contracts and ensuring their compliance with legal standards. By improving the quality of legal documentation, this strategy seeks to protect the interests of all parties to a contract.

Keywords: Legal contract · Bayesian Network · Risk Assessment

1 Introduction

In the legal field, contractual errors can have significant consequences for the parties involved. Errors relating to contractual terms or the identification of the parties can lead to a range of legal consequences, requiring in-depth analysis to determine the appropriate remedial action. In addition, errors relating to the factual and legal assumptions underlying the contract can compromise its validity and performance, which may require a thorough reassessment of the contractual provisions. Incomplete contracts also present a challenge, as they are generally considered void in their current form. However, in some cases, it is possible to remedy these deficiencies to render the contract valid and enforceable, thus preserving its integrity and legality [2]. In the event of partial invalidity of the contract, the courts have several options for remedying the situation, such as annulling the entire contract or enforcing the remaining valid provisions. This decision depends on various factors, including the nature and extent of the invalidity, as well as the implications for the parties involved [3]. At the same time, the application of artificial intelligence and automation systems to contracts is transforming the way contracts are created, managed and analysed, with a significant impact on the legal market and professionals in the field. It is crucial to underline the importance of compliance with legal standards in contract content with this digitisation.

In this article, we underscore the utilization of Bayesian networks for modeling the risk of contract invalidity. Bayesian networks, potent tools for probabilistic analysis, can graphically represent the dependencies among different variables impacting the validity of a contract [8, 9], thereby facilitating the identification of key risk factors. By amalgamating empirical data and expert knowledge [10], these networks can also estimate the probability of various contract invalidity scenarios occurring. The aim of this approach is to evaluate and anticipate factors that could undermine the validity of a contract, whether through ambiguities in contractual terms, non-compliance with legal requirements, or other deficiencies in consent. The validity of a contract hinges on a multifaceted array of factors, spanning from adherence to applicable laws and regulations to the clarity and precision of contractual terms. Consequently, modeling the risk of contract invalidity entails a comprehensive analysis of these diverse elements to identify potential vulnerabilities and implement suitable preventive measures. This methodology offers a proactive approach to managing the risks associated with contract validity, enabling parties to make informed decisions to avert disputes and ensure contract compliance with the law.

2 The Basics of Legal Contract Formation

The formation of a legally binding contract is based on several fundamental components that ensure its validity and performance in accordance with established legal standards. These elements are as follows:

- **Offer, Acceptance and Consideration:** The formation of a valid contract generally requires the existence of a clear and precise offer, the unreserved acceptance of that offer and an equivalent consideration in exchange for the contractual obligations [11–13].
- **Legal requirements:** To be valid, a contract must satisfy certain legal requirements. It must be clear and certain in its terms, involve mutual consideration, reflect the intention of the parties to create legal relations, and comply with the principles of illegality and public policy [12].
- **Structural complexity:** Contracts often have a complex structure and the way in which they are drafted can affect their interpretation [15, 16].

3 Related Work

Assessing and enhancing the quality of contracts represents a critical domain of inquiry within both management and law. Valuable recommendations have emerged to enhance the quality of contractual documentation, emphasizing the significance of lucid descriptions, avoidance of implicit assumptions, and proactive drafting to mitigate flaws and potential disputes. Furthermore, research indicates that amalgamations of relational and associative contractual clauses are associated with heightened operational performance, shedding light on the strategic deployment of contractual clauses to optimize operational effectiveness. Within the realm of construction contracts specifically, endeavors have been undertaken to assess and enhance contract quality assurance. Studies have employed the Analytical Hierarchy Process (AHP) to juxtapose contract quality across different contractors [17], while others have scrutinized the obligations of proprietors under FIDIC construction contracts, identifying pivotal obligations that, if comprehended adequately, can curtail disputes and their attendant costs [18]. Technological advancements have further reshaped the contractual landscape, with the burgeoning integration of artificial intelligence (AI) and automation. This evolution transcends contracts from mere documents to data managed by technology, facilitating streamlined interactions among involved parties [5]. To appraise the quality of smart contracts, Bayesian network models have been devised, demonstrating a notable success rate in this regard [8]. Additionally, Bayesian networks have been leveraged to assess diverse strategies for mitigating contractual risks, including within aerospace supply chains [19], and to delineate risk management processes, offering promising avenues for large-scale mitigation of contractual risks. This body of research underscores the escalating importance of technological innovations and analytical methodologies in enhancing contract quality and management, heralding more effective and efficient legal and commercial practices [9, 20].

4 Methodology

4.1 Overview of the Bayesian Network

Bayesian networks, a class of models widely used in artificial intelligence and decision support systems [21, 22], provide a formal and rigorous framework for understanding reasoning under uncertainty. Essentially, these networks describe the relationships between variables using a directed acyclic graph [23], in which the nodes represent the variables and the directed arcs reflect the probabilistic dependencies between these variables. This graphical representation provides an intuitive visualisation of the complex interrelationships between the different components of the system being modelled, allowing in-depth analysis of the underlying mechanisms [23] (Fig. 1).

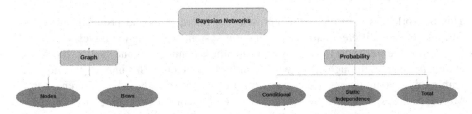

Fig. 1. Visualizing the Framework of Bayesian Networks

A Bayesian network comprises a graph structure and conditional probability tables (CPTs), which reflect the strength of causality between exogenous and endogenous variables as well as the likelihood of events occurring [24]. The network's structure is defined by nodes and directed edges [25], with the mathematical interpretation provided by conditional probabilities. This method enables the tackling of practical problems from both qualitative and quantitative perspectives [23].

$$BN_R = (C_i, P) \tag{1}$$

Equation (1) expresses this relationship such that BN_R represents the configuration of the Bayesian network, while P denotes the set of directed edges, and C_i includes the set of nodes that make up the network [23, 24].

$$P(C_i|C_j, C_n) = P(C_i|C_j) \qquad (2)$$

On the other hand, the structure of a Bayesian network is characterized by the assumption of conditional independence, which means that each node is conditionally independent of nodes that are not its descendants, as long as the parent node is identified [23]. Equation (2) represents this assumption. In this formulation, Cj represents the parent node of Ci, while P(Cj) denotes the probability of events associated with the parent node. As for Cn, it includes the set of child nodes other than Ci.

4.2 Modelling the Invalidity Risk of a Legal Contract

This article proposes a risk assessment model for the formation of legal contracts using the causal structure of Bayesian networks, thus providing a quantitative assessment of the risk associated with this process, based on practical experience and fundamental theoretical knowledge. In the first phase, the risk factors associated with the formation of legal and quality contracts were identified through expert interviews and an in-depth analysis of the literature. Next, the causal logic linking these risk factors was presented in the form of a causal diagram and then mapped onto the Bayesian network model. Next, the conditional probability of the endogenous variables was determined based on feedback from lawyers and legal professionals collected through a questionnaire. Finally, once the risk assessment was complete, the Bayesian network solution for creating legally valid, high-quality contracts was obtained, allowing the level of risk associated with the process to be calculated. An in-depth analysis of the critical influencing factors was also carried out. The different stages of this approach are illustrated in Fig. 2.

4.3 Developing the Causal Graph

This network was constructed to represent the interactions and probabilistic dependencies between different variables within our system. By using the Bayesian network framework, we were able to visually represent these connections and analyze how information propagates through the network to influence outcomes. It's important to note that within this system, the legal accuracy aspect of a contract is based on references from Moroccan law, specifically the legislation governing contract management in Morocco, commonly referred to as the 'Code of Obligations and Contracts'. Consequently, compliance with the contract within this framework is harmonised with the applicable Moroccan legislation.

The Bayesian network of our system depicted in Fig. 3 illustrates the causal relationships between various parameters, each variable has been qualitatively represented by expressions in natural language, as illustrated in Table 3, the network organized into three levels:

- The first level describes the input nodes, which indirectly impact the risk of contract invalidity. These nodes, categorized into six groups as demonstrated in Table 2 constitute the initial level.

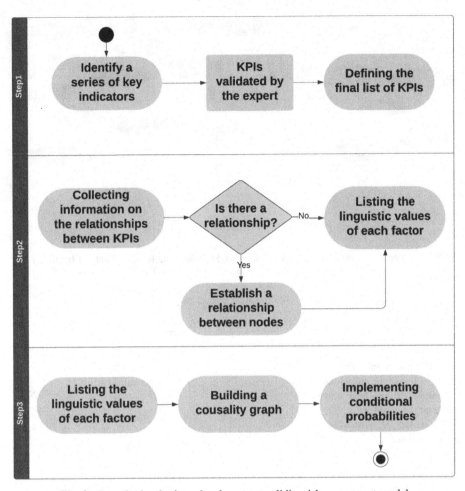

Fig. 2. Steps in developing a legal contract validity risk assessment model

- The second level consists of intermediate nodes that lead to final impacts., as detailed in Table 1.
- At the third level are the final effects directly influencing the contract's validity and legality

4.4 Generating Conditional Probabilities

The implementation of a Bayesian network can be approached through various methods, each presenting distinct advantages and drawbacks. Objective methods utilize machine learning algorithms to derive the network structure from available datasets. While these techniques are often effective for extensive datasets, they may demand substantial computational resources and may not be ideal for scenarios with limited data availability [24]. In contrast, subjective methods involve soliciting insights from domain experts

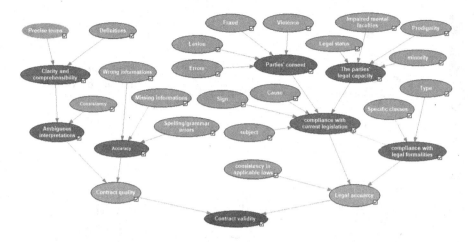

Fig. 3. The Structure of the Bayesian Network Modeling the Risk of Contract Invalidity

Table 1. Intermediate effects and impacts

Intermediate Effect	Impacts
• Accuracy • Clarity and comprehensibility • Ambiguous interpretations • compliance with legal formalities • the parties' consent • The parties' legal capacity • compliance with current legislation • specific formalities	• Legal accuracy • Contract quality • Contract validity

through interviews, surveys, or other qualitative data collection techniques. For this study, the subjective method was chosen due to its flexibility and its capacity to integrate expert knowledge explicitly. Unlike objective approaches, which rely heavily on extensive data, subjective methods offer greater adaptability to specific contexts and enable more accurate integration of qualitative expertise [21, 23].

To construct the conditional probability tables (CPTs) within our system, we devised a questionnaire (refer to Fig. 4) aimed at an appropriate sample of experienced professionals specializing in contract analysis [10, 23]. This sample includes lawyers and contract management advisors. Drawing on their practical expertise in evaluating and drafting contracts, these individuals possess the necessary skills to provide valuable insights into the quality and legality of contracts, as well as any associated risks [5].

This questionnaire (Fig. 4) is designed to qualitatively assess the compliance of the contract in terms of legal accuracy, by collecting the states of various factors that have a significant impact on the node. It uses scales to facilitate the collection of information from experts, Table 4 explains the meaning of each scale.

Table 2. Description of the causal graph input parameters

Variable Class	Variable name	Description
Clarity [29–32]	Precise terms	Will help to make the information easier to understand and to make it clear and unambiguous [33]
	Definitions	
Accuracy [31, 32]	Wrong information [30]	Content free from omissions and errors [29, 31, 32]
	Spelling/grammar errors [16]	
	Missing information [29]	The document must contain all the essential data [29, 33]
Conformity [16, 31]	Consistency in applicable laws	The coherence of laws in different legal systems
	Specific clauses [34]	The clauses serves to define the scope, obligations, and rights specific to the contract's nature [33]
	Type [34]	
Interpreting [26]	Consistency [16, 31]	Coherent and harmonious drafting for relevant coordination between the various disciplines [33]
Capacity	Legal status [27]	Refers to the capacity of an entity, whether a natural person (individual) or a legal person (company), to possess rights and obligations and to take legal action [26, 28]
	Impaired mental faculties [27]	Mental incapacity, they are unable to make informed decisions or consent to legal acts [28]
	Prodigality [27]	Legal concept for individuals unable to manage affairs due to excessive spending or reckless finances [26, 27]
	minority [27]	the age at which an individual is recognized as an adult by law [28]
Parties' consent	Lesion	Damage resulting from unequal reciprocal benefits in a contract [26, 28]
	Violence [26, 27]	Moral or physical harassment [26]
	Fraud [26, 27]	One of the contracting parties has been the victim of deception on the part of the other party [28]

(*continued*)

Table 2. (*continued*)

Variable Class	Variable name	Description
Content [26]	Errors [26, 27]	An erroneous and inaccurate conception held by the contractor [27]
	Subject [26, 28]	Each party has the option of undertaking to carry out or not to carry out a specific action, subject to compliance with the law and standards of decency [26]
	Cause [26, 28]	Refers to the reason why each party has agreed to enter into the agreement [26]
	Sign [27]	A sign, whether handwritten or electronic, is the act by which a person expresses his or her consent, agreement or commitment to the content of a contractual document [26]

Table 3. Linguistic values of nodes

Nodes	Linguistic values
Precise terms	Good, Medium, Bad
Definitions	Present, Absent
Wrong information	High, Medium, Low
Spelling/grammar errors	High, Medium, Low
Missing information	High, Medium, Low
Consistency in applicable laws	Present, Absent
Specific clauses	Present, Absent
Type	Yes, No
Consistency	Good, Medium, Bad
Legal status	Present, Absent
Impaired mental faculties	Present, Absent
Prodigality	Present, Absent
minority	Major, Emancipated, Minor
Lesion	Present, Absent
Violence	Present, Absent
Fraud	Present, Absent

(*continued*)

Table 3. (*continued*)

Nodes	Linguistic values
Errors	High, Medium, Low
Subject	Legal, Illegal
Cause	Legal, Illegal
Sign	Present, Absent
Accuracy	High, Medium, Low
Clarity and comprehensibility	Clair, Medium, Obscure
Ambiguous interpretation	High, Medium, Low
Compliance with legal formalities	Present, Absent
The parties' consent	Present, Absent
The parties' legal capacity	Full capacity, Limited, Incapacity
Specific formalities	Present, Absent
Legal accuracy	Present, Uncompleted, Absent
Contract quality	Good, Intermediate, Bad
Contract validity	Valid, Relative invalidity, Absolute invalidity

5   Results and Discussion

The Bayesian network we developed highlights three states that characterize the validity of a contract: absolute invalidity, relative invalidity, and a valid contract under Moroccan law. Absolute invalidity occurs when a contract violates a mandatory rule of law that cannot be circumvented. These different states largely depend on the degree of compliance with the intermediate node called "legal precision' in our network, as illustrated in Fig. 5. To analyze the factors influencing this node, we pay particular attention to those that have a significant impact on legal precision and, consequently, on the validity of the contract. By highlighting these elements, we can identify the factors likely to lead to the absolute nullity or validity of the contract under Moroccan law.

In the second part, the relative invalidity of a contract depends largely on the quality of the contract in terms of ambiguity, clarity and omissions or errors. This part of our network affects two main aspects. Firstly, it affects the accuracy of the parties' consent under the law. Secondly, it has a direct impact on the validity of the contract, with the aim of guaranteeing the quality of the document and preventing it from being considered null and void under the law.

Analysis of the results from the simulation of the different scenarios in our Bayesian network, as illustrated in Table 5, highlights several significant trends. Firstly, when looking at contract quality, we observe a clear correlation between perceived 'bad' quality and higher rates of relative and absolute disability. Scenarios 1, 2 and 3, where policy quality is rated as 'bad', have higher relative and absolute disability rates than the other scenarios. With regard to legal accuracy, the results show that the presence of higher legal accuracy, represented by the fact that 'legal accuracy' is classified as

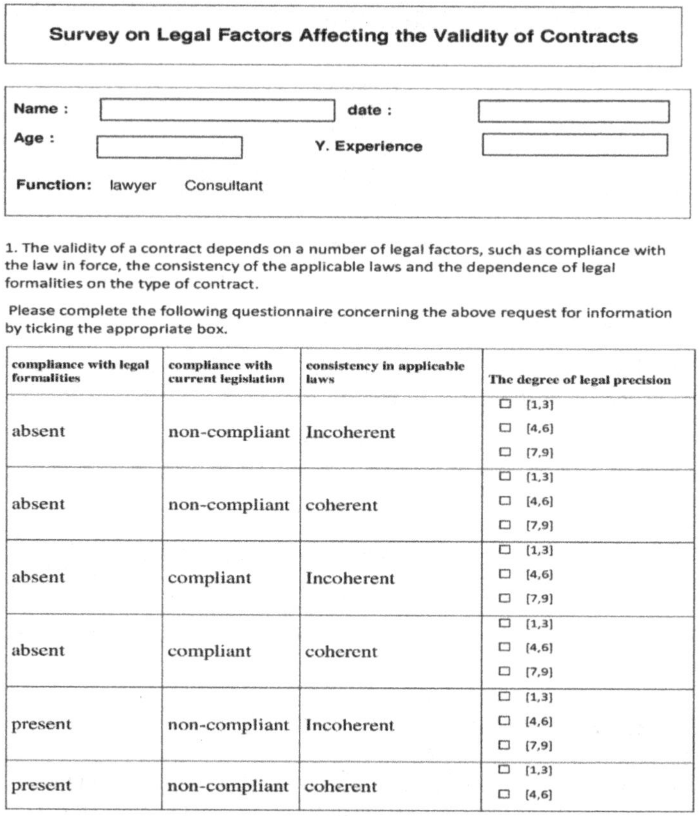

Survey on Legal Factors Affecting the Validity of Contracts

Name: ☐ date: ☐
Age: ☐ Y. Experience ☐
Function: lawyer Consultant

1. The validity of a contract depends on a number of legal factors, such as compliance with the law in force, the consistency of the applicable laws and the dependence of legal formalities on the type of contract.

Please complete the following questionnaire concerning the above request for information by ticking the appropriate box.

compliance with legal formalities	compliance with current legislation	consistency in applicable laws	The degree of legal precision
absent	non-compliant	Incoherent	☐ [1,3] ☐ [4,6] ☐ [7,9]
absent	non-compliant	coherent	☐ [1,3] ☐ [4,6] ☐ [7,9]
absent	compliant	Incoherent	☐ [1,3] ☐ [4,6] ☐ [7,9]
absent	compliant	coherent	☐ [1,3] ☐ [4,6] ☐ [7,9]
present	non-compliant	Incoherent	☐ [1,3] ☐ [4,6] ☐ [7,9]
present	non-compliant	coherent	☐ [1,3] ☐ [4,6]

Fig. 4. Example of a legal agreement evaluation questionnaire

Table 4. Correspondence between qualitative and quantitative intervals

Scale	Quantitative value	Qualitative value
[1–3]	20%	Absent
[4–6]	50%	Incomplete
[7–9]	80%	Present

'present', is generally associated with lower relative and absolute invalidity rates. For example, scenario 2, where legal accuracy is classified as 'present', has lower relative and absolute disability rates than scenario 1 where legal accuracy is classified as 'absent'. Finally, with regard to contract completion, represented by the 'Uncompleted' status, the results show higher relative and absolute disability rates. For example, scenarios 3 and 6, where the contract is not completed, show higher relative and absolute disability rates than the other scenarios.

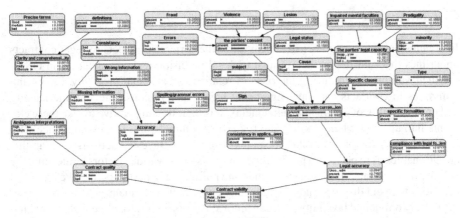

Fig. 5. Legal contract validity measurement

Table 5. Correspondence between qualitative and quantitative intervals

Scenarios	Contract quality	Legal Accuracy	Contract validity probability		
			Valid	Relative invalidity	Absolute invalidity
Scenario 1	Bad 0.61	Absent 0.79	0.11	0.07	0.82
Scenario 2	Bad 0.46	Present 0.88	0.62	0.31	0.07
Scenario 3	Bad 0.47	Uncompleted 0.58	0.32	0.13	0.55
Scenario 4	Intermediate 0.43	Absent 0.54	0.27	0.16	0.57
Scenario 5	Intermediate 0.44	Present 0.49	0.39	0.21	0.38
Scenario 6	Intermediate 0.41	Uncompleted 0.57	0.32	0.16	0.52
Scenario 7	Good 0.69	Absent 0.91	0.03	0.04	0.93
Scenario 7	Good 0.81	Present 0.88	0.84	0.09	0.06
Scenario 8	Good 0.72	Uncompleted 0.49	0.28	0.17	0.55

In summary, these results highlight the importance of contract quality, legal precision, and contract completion in determining contract validity and vulnerability to relative and absolute invalidity. Factors influencing contract quality and legal accuracy must be taken into account when drafting and reviewing contracts to ensure their validity and reduce the risk of invalidity, as shown in Fig. 5. By ensuring that contracts are drafted clearly, precisely and in accordance with legal standards, the parties involved can minimise the chances of relative or absolute invalidity, and thus protect their interests more effectively.

6 Conclusion

In conclusion, this comprehensive analysis of the formative aspects of legal contracts underlines the vital importance of quality, legal precision and completeness in ensuring their validity and reducing the risks of relative and absolute invalidity. Errors and omissions in contracts can have significant legal repercussions, affecting the legality and enforceability of contractual agreements. It is therefore imperative to take these elements into account when drafting and reviewing contracts, to ensure compliance with legal standards and prevent future disputes. The use of Bayesian network tools in this article enables us to perform a probabilistic analysis of the various factors influencing the validity of a contract, thus offering contracting parties the opportunity to make informed decisions to avoid disputes and ensure legally compliant performance. Thus, modelling the risk of contractual invalidity offers an innovative perspective for effectively managing the risks associated with contract validity, thereby reinforcing confidence in contractual agreements and ensuring better protection of the interests of the parties involved.

References

1. Ghaly, K.: Mistake, misrepresentation, and frustration (2021). https://doi.org/10.1093/oso/9780198832805.003.0008
2. Ramos, C.P.: The incomplete contract | EL negocio incompleto. Revista de Derecho Civ. **10**(5), 43–91 (2023)
3. Tomás, G.: Utile per inutile non vitiatur: can favor contractus be considered a European regula iuris? Eur. Rev. Contract Law **12**(3), 259–286 (2016). https://doi.org/10.1515/ercl-2016-0014
4. Martinelli, S.: AI as a tool to manage contracts consequences and challenges in applying legal tech to contracts management. Eur. Rev. Priv. Law **31**(2–3) (2023)
5. Martinelli, S.: AI as a tool to manage contracts consequences and challenges in applying legal tech to contracts management. Eur. Rev. Priv. Law **31**(2–3), 1–16 (2023)
6. Pelikánová, I.: Determination of a contractual relationship independent of the will of the parties. | Determinace smluvního vztahu nezávislá na vůli stran. Pravny Obzor **106**(5), 392–415 (2023). https://doi.org/10.31577/pravnyobzor.2023.5.02
7. Mihai-Adrian, B.S., et al.: Improving safety in the workplace using checklists legal requirements. Qual. Access Success **18** (2017)
8. Sathiyamurthy, K., Kodavali, L.: Bayesian network-based quality assessment of blockchain smart contracts (2023). https://doi.org/10.1016/bs.adcom.2023.07.004
9. Liu, Q., Tchangani, A., Pérès, F., Gonzalez-Prida, V.: Object-oriented Bayesian network for complex system risk assessment. Proc. Inst. Mech. Eng. O J. Risk. Reliab. **232**(4), 340–351 (2018). https://doi.org/10.1177/1748006X17753026
10. Timmer, S.T., Meyer, J.J.C., Prakken, H., Renooij, S., Verheij, B.: Explaining Bayesian networks using argumentation. In: Destercke, S., Denoeux, T. (eds.) ECSQARU 2015. LNCS, vol. 9161, pp. 83–92. Springer, Cham (2015). https://doi.org/10.1007/978-3-319-20807-7_8
11. Marson, J., Ferris, K.: 7. Factors affecting the validity of a contract. Bus. Law (2020). https://doi.org/10.1093/he/9780198849957.003.0007
12. Schroeter, U.G.: Contract validity and the CISG. Uniform Law Rev. **22**(1) (2017). https://doi.org/10.1093/ulr/unx010
13. Abdul-Malak, M.-A.U., Ezzeddine, F.: Contract documents defects: ensuring contract drafters know what they want and say it clearly. J. Legal Aff. Dispute Resolut. Eng. Constr. **15**(3) (2023). https://doi.org/10.1061/JLADAH.LADR-927

14. Verstappen, J.: Formation of contracts. In: Law, Governance and Technology Series, vol. 56 (2023). https://doi.org/10.1007/978-3-031-35407-6_3
15. Hwang, C., Jennejohn, M.: Deal structure. Northwest Univ Law Rev. **113**(2) (2018). https://doi.org/10.2139/ssrn.3043860
16. Andi, Minato, T.: Design documents quality in the Japanese construction industry: factors influencing and impacts on construction process. Int. J. Proj. Manag. **21**(7) (2003). https://doi.org/10.1016/S0263-7863(02)00083-2
17. Vilkonis, A., Antucheviciene, J., Kutut, V.: Construction contracts quality assessment from the point of view of contractor and customer. Buildings **13**(5) (2023). https://doi.org/10.3390/buildings13051154
18. Alhyari, O.: Owners' obligations under FIDIC construction contracts (2021). https://doi.org/10.1007/978-3-030-48465-1_33
19. Qazi, A., Quigley, J., Dickson, A., Gaudenzi, B., Ekici, S.O.: Cost and benefit analysis of supplier risk mitigation in an aerospace supply chain. In: Proceedings of 2015 International Conference on Industrial Engineering and Systems Management, IEEE IESM 2015, pp. 850–857 (2016). https://doi.org/10.1109/IESM.2015.7380255
20. Liu, Q., Pérès, F., Tchangani, A.: Object oriented Bayesian network for complex system risk assessment. IFAC-PapersOnLine, 31–36 (2016). https://doi.org/10.1016/j.ifacol.2016.11.006
21. Wiegerinck, W., Burgers, W., Kappen, B.: Bayesian networks, introduction and practical applications. In: Bianchini, M., Maggini, M., Jain, L. (eds.) Handbook on Neural Information Processing. Intelligent Systems Reference Library, vol. 49, pp. 401–431. Springer, Heidelberg (2013). https://doi.org/10.1007/978-3-642-36657-4_12
22. Stefanini, F.M.: Chain graph models to elicit the structure of a Bayesian network. Sci. World J. **2014** (2014). https://doi.org/10.1155/2014/749150
23. Rao, M.B., Rao, C.R.: Bayesian networks, vol. 32 (2014). https://doi.org/10.1016/B978-0-444-63431-3.00010-3
24. Cooper, G.F., Herskovits, E.: A Bayesian method for the induction of probabilistic networks from data. Mach. Learn. **9**(4) (1992). https://doi.org/10.1007/bf00994110
25. Wang, Y., Su, J., Zhang, S., Guo, S., Zhang, P., Du, M.: A dynamic risk assessment method for deep-buried tunnels based on a Bayesian network. Geofluids **2020** (2020), https://doi.org/10.1155/2020/8848860
26. Corinne: Renault-Brahinsky, L'essentiel du droit des obligations
27. Lex, W.: Code des obligations et des contrats, notamment les articles 77, 79, 80, 81, 84, 264 et 435, (Dahir du 12 août 1913 (9 ramadan 1331)) (1913)
28. Les obligations formées par l'acte juridique-le contrat
29. Tang, S.L., Lu, M., Chan, Y.L.: Achieving client satisfaction for engineering consulting firms. J. Manag. Eng. **19**(4) (2003). https://doi.org/10.1061/(asce)0742-597x(2003)19:4(166)
30. Samson, D., Parker, R.: Service quality: the gap in the Australian consulting engineering industry. Int. J. Qual. Reliab. Manag. **11**(7) (1994). https://doi.org/10.1108/02656719410738993
31. Tilley, P.A., Mcfallan, S.L., Tucker, S.N.: Design and documentation quality and its impact on the construction process. Constr. Eng. (1999)
32. Aluko, O.R., Idoro, G.I., Mewomo, M.C.: Relationship between perceived service quality and client satisfaction indicators of engineering consultancy services in building projects. J. Eng. Des. Technol. **19**(2) (2021). https://doi.org/10.1108/JEDT-03-2020-0084
33. Govender, N., Laryea, S., Watermeyer, R.: A framework for assessing quality of tender documents. Built Environ. Proj. Asset Manag. **12**(4), 573–589 (2022). https://doi.org/10.1108/BEPAM-07-2021-0094
34. D. (19. .-. . . . ; juriste). Boustani, L'essentiel du droit des contrats spéciaux

Optimizing Recommendation Systems in E-Learning: Synergistic Integration of Lang Chain, GPT Models, and Retrieval Augmented Generation (RAG)

Qamar EL Maazouzi[✉], Asmaâ Retbi, and Samir Bennani

RIME TEAM-Networking, Modeling and e-Learning Team, MASI Laboratory, Engineering, 3S Research Center, Mohammadia School of Engineers, Mohammed V University, Rabat, Morocco
qamarelmaazouzii65@gmail.com, sbennani@emi.ac.ma

Abstract. The emergence of recommendation systems stands as an essential response to the increasing complexity of choices, providing personalized suggestions. Recent advances in artificial intelligence, particularly in natural language processing (NLP), enhance the effectiveness of these systems, driven by cutting-edge models such as OpenAI's GPT-3.5. However, challenges persist, including the "cold start" and hallucination issues within Large Language Models (LLMs). Our research focuses on exploring the use of OpenAI's embedding to enhance online course recommendations, while integrating LLMs Lang Chain and GPT-3.5 into the E-learning domain. We introduce the "Retrieval Augmented Generation" (RAG) approach to address these challenges. Experiments conducted on a real E-learning dataset evaluate the models' ability to understand and generate relevant recommendations. The results demonstrate the effectiveness of the contributions, emphasizing the specific opportunities provided by the RAG approach to tackle challenges in recommendation systems based on LLMs.

Keywords: Recommendation Systems · Natural Language Processing (NLP) · Large Language Models (LLMs) · E-learning · Retrieval Augmented Generation (RAG) · GPT

1 Introduction

The proliferation of data, driven by the increasing digitization of information and the expansion of social media, generates a wealth of information about user behaviors and preferences. With the diversification of services, from streaming

---

A. Retbi and S. Bennani—These authors contributed equally to this work.

platforms to e-commerce sites, users face a multitude of choices. In this environment, recommendation systems emerge as essential tools to simplify this complexity, offering relevant and personalized suggestions, thereby enhancing the user experience.

The dynamic evolution of recommendation systems has garnered growing interest in the research community. This progress stems from the integration of more advanced methodologies, including collaborative filtering models, as mentioned by [1], which leverages disease types and hospital locations to assist patients in choosing hospitals. Constraints of the model, related to data scarcity and the cold start challenge, have led to the adoption of more advanced approaches, particularly those based on deep learning [2]. Another method, content-based filtering introduced by [3], generates personalized movie recommendations based on textual features of the films. A hybrid approach, often combining multiple techniques, has been highlighted by [4]. On the other hand, Knowledge-based Recommendation Systems, explained by [5], use explicit knowledge about users and items to generate recommendations. Additionally, the advent of artificial intelligence and advancements in natural language processing (NLP) have played a crucial role.

Advanced models like OpenAI's GPT-3.5 demonstrate a finer understanding of language, enabling recommendation systems to better interpret user needs and preferences. Various studies have been conducted to leverage recent advances in natural language processing (NLP) and the use of Large Language Models (LLMs). [6] provides a review of LLM-powered recommendation systems, covering aspects such as pre-training, fine-tuning, and prompting. Meanwhile, [7], in their study, explores and evaluates the potential of ChatGPT in the recommendation domain, leveraging its extensive linguistic and global knowledge acquired from large-scale corpora.

However, despite these advancements, recommendation systems (RS) face challenges such as the "cold start," occurring when RS fail to offer relevant suggestions for new users or items with limited or nonexistent historical data. Furthermore, Large Language Models (LLMs) may be subject to hallucination constraints, resulting in the generation of fictitious or incorrect information. These limitations often lead to unreliable responses, casting doubt on the effectiveness of these systems in dynamic environments.

To address these issues, our research focuses on the convergence of recommendation systems, natural language processing (NLP), and the "Lang Chain" method. The goal is to explore potential synergies to provide more accurate and adaptive recommendations. To guide our research, we formulate the following questions:

1. (R1): How can OpenAI's embedding be utilized for improved online course recommendations?
2. (R2): How can LLMs, specifically Lang Chain and GPT 3-5, be used to address recommendation system tasks in the E-learning domain?

3. (R3): How to introduce the "Retrieval Augmented Generation" (RAG) approach with Lang Chain and embedding to address the challenges of cold start and hallucination for LLMs?

To answer these questions, various experiments were conducted, emphasizing key research points. Initially, experiments were conducted to explore how Large Language Models (LLMs), including Lang Chain and GPT-3.5, can be integrated effectively into recommendation systems for E-learning. This includes evaluating these models' ability to understand and generate relevant recommendations based on learner behaviors and preferences.

All these experiments were conducted on a real dataset from the E-learning domain. The results were compared with other existing recommendation methods to provide a comprehensive assessment of the effectiveness of the contributions.

In summary, our article aims to address research questions by making significant contributions to the improvement of natural language-based recommendation systems, with a focus on our E-learning research context. In this context, our research approach provides the following contributions:

1. Exploration of LLMs' Potentials: We examine how to leverage the potential of Large Language Models (LLMs), particularly Lang Chain and GPT-3.5, to substantially optimize educational recommendations.
2. Highlighting the "Retrieval Augmented Generation" (RAG) Approach: Our study focuses on distinct possibilities linked to employing the "Retrieval Augmented Generation" (RAG) approach. This strategy is specifically emphasized as a remedy for tackling the issues of cold start and hallucination encountered by Large Language Models (LLMs)

The remainder of the article is structured into four sections. Section 2 delves into the theoretical framework, highlighting the use of the "Retrieval Augmented Generation" (RAG) approach and language models. Section 3 provides a literature review, pinpointing existing gaps. Section 4 presents the research results and initiates a discussion on their implications. Section 5 concludes the article by summarizing key findings and suggesting directions for future research in the field of recommendation systems based on large language models.

2 Theoretical Background

2.1 Recommendation Systems

Recommendation systems are software applications that utilize advanced algorithms and techniques to analyze user behavior, preferences, and past interactions in order to provide personalized suggestions [8]. These systems aim to anticipate the needs or preferences of users by recommending products, services, or content that align with their individual profiles. There are two main types of recommendation systems: content-based, which use information about items or users to make suggestions, and collaborative, which rely on the past behaviors and preferences of similar users.

2.2 Large Language Models (LLM) and GPT

LLM (Large Language Model) refers to a language model capable of processing and generating text on a large scale using machine learning techniques. These models are trained on extensive sets of textual data to learn grammatical structures, semantic context, and linguistic relationships.

GPT (Generative Pre-trained Transformer) is a specific example of an LLM. GPT is a family of models developed by OpenAI that utilizes a transformer architecture to perform natural language processing tasks. These models are pre-trained on massive datasets, enabling them to understand and generate text contextually and fluently. GPT-3, for instance, is one of the most powerful GPT models, with 175 billion parameters [9].

2.3 Langchain

LangChain represents an innovative framework dedicated to the simplified development of a variety of applications leveraging generative AI language models. Whether it's for creating chatbots, personal assistants, summarization tools, or question/answer generation, LangChain provides a versatile modular solution. In addition to these applications, LangChain offers extended features for code writing and comprehension, as well as interaction with APIs, thereby paving the way for a multitude of other use cases.

2.4 Embedding

Embedding, a crucial technique in Natural Language Processing (NLP), involves representing elements such as words or phrases in vector form within a multidimensional space [10]. This vector representation provides an efficient way to capture semantic and contextual relationships between elements, There by facilitating their manipulation and analysis. There are various types of embeddings, each suited to specific tasks. Among OpenAI's embedding models, we can mention ADDA 2 (Adversarial Discriminative Domain Adaptation), which has transformed the way NLP is represented. These embeddings are particularly important in the context of transfer learning, enabling the application of understanding from one specific domain to another.

2.5 Retrieval Model

The Retrieval-Augmented Generation (RAG) model is an advanced approach in the field of Natural Language Processing (NLP) that combines two main components: the Retrieval Model and the Generative Model [11].

1. Retrieval Model: Retrieval models extract information from the knowledge base using semantic search techniques, often involving the use of embeddings. These models measure the semantic similarity between the user's query and the embeddings associated with documents in the knowledge base. Using this similarity, retrieval models identify and select the most relevant passages or information in the knowledge base.

2. Generative Model: The generative model uses the information extracted by the retrieval models as context to generate text in a contextually appropriate manner. Language Modeling Models (LLMs), such as GPT (Generative Pre-trained Transformer), are frequently employed for this component, leveraging pre-trained models capable of generating coherent and informative text.

2.6 Vector Database

A vector database is a specific data structure that stores and manages multidimensional vectors, often representing data in a mathematical space with multiple dimensions. It provides advanced functionality for conducting search, similarity, and analysis operations on the stored vectors. These databases are particularly valuable in the fields of machine learning, natural language processing, and recommendation systems, where the efficiency of operations on vectors is crucial. Various types of vector databases are available, each with specific characteristics and advantages. Among them, Chroma and Pinecone stand out due to their performance and adaptability in handling embedding vectors.

3 Previous Work

Several research studies have been conducted on recommendation systems across various domains, employing diverse approaches and models. In [12], the author addresses the challenge of movie selection by leveraging recommendation systems through the use of data analysis, machine learning algorithms, and temporal analysis techniques. These systems incorporate user behaviors to provide accurate recommendations. Collaborative filtering algorithms identify similarities between users or movies, while content-based filtering segments movie features based on user preferences. The application of time series analysis methods captures temporal patterns for dynamic recommendations, adapting to changes in information and preferences, thereby offering relevant suggestions. However, it is essential to consider the impacts of real-time data collection and the reliability of the hybrid system.

In the same domain, the article [13] adopts an ensemble recommendation system model based on RNN, combining the results of four different individual learners. The cognitive approach is integrated by considering the user's age, recommending genres based on age group psychology, recognizing the latent influence of age and region on genre preferences.

In another domain related to e-commerce, the authors in [14] propose a personalized recommendation system by integrating user data from social networks and e-commerce. It emphasizes user profile creation using textual information, tags, and relationships. The use of a topic model and collaborative filtering algorithms improves the accuracy of recommendations, with a feedback mechanism. Empirical analysis confirms the impact of social networks on e-commerce. However, the article does not specify the size or representativeness of the dataset used and does not compare the system with other models to assess its performance.

Regarding the focus of our research, various experiments have been conducted on recommendation systems specifically designed for online learning platforms. Thus, the research in [15] presents a Recommendation System using Content-Based Filtering with Jaccard and Cosine Similarity algorithms, demonstrating their effectiveness according to the RMSE metric. However, the research is limited to specific algorithms and a restricted dataset, the Coursera Free Dataset, potentially impacting the generalizability of the results. Additionally, the evaluation solely focuses on the RMSE metric, neglecting other indicators that could enrich the system analysis. In the same context, the article [16] introduces a recommendation system for online courses using collaborative filtering models (KNN, SVD, and NCF), evaluated through different metrics. The results indicate that KNN outperforms other models. However, the author limits the evaluation to collaborative filtering models, neglecting to explore other recommendation algorithms or hybrid approaches. Moreover, the study uses a specific dataset from Coursera, which may not be representative of all online learning platforms.

Considering this perspective, the study [17] aligns with the broader context of exploring recommendation systems. The article explores the use of Large Language Models (LLMs) such as GPT-4. It formulates the problem as a conditional ranking task, with experiments demonstrating interesting ranking capabilities of LLMs. The results suggest competitiveness with conventional recommendation models. However, challenges such as the perception of historical interactions and position biases are identified. On the other hand, the author in [18] provides an analysis of user rating prediction by comparing the performance of Collaborative Filtering (CF) and Large Language Models (LLMs) in zero-shot, few-shot, and fine-tuning scenarios. By evaluating various LLMs of varied sizes, from 250M to 540B parameters, it highlights their effectiveness. However, the research primarily focuses on the comparison with CF, neglecting to explore other recommendation models or compare with a broader range of references. While demonstrating the potential of LLMs in terms of data efficiency, the study exclusively focuses on predicting user ratings, leaving other aspects of the recommendation system underexplored

4 Methodology

4.1 Embedding Open AI

To address our first research question (R1), we introduce our initial conceptual framework that focuses on embedding using the OpenAI Text Embedding Ada-002 model (Fig. 1). The architecture relies on cosine similarity, leveraging the use of the vector database "Pinecone." The workflow for generating recommendations is detailed below (Fig. 1):

1. Embedding Generation with ADDA-2: After downloading our course dataset, we use the embedding model.
2. User Input:
   (i) The user provides a query for a course.

(ii) The user's query is converted into an embedding using the "text-embedding-ada-002" model.

(iii) The user's embedding is normalized.

3. Indexing in Pinecone: We store these embeddings in a vector database (Pinecone).
4. Pinecone Database and Cosine Similarity: The vector database (Pinecone) receives the embedding (the query) and applies a similarity search algorithm (cosine similarity) to find the indexed embeddings most similar to the given query.
5. Nearest Neighbors Results: The results of the nearest neighbors from the vector database are organized based on the relevance/similarity score, where the top-ranked skills are returned as search results for the given query in JSON format.

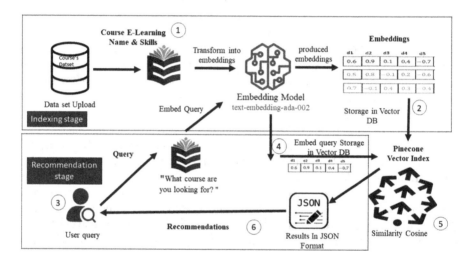

Fig. 1. Framework for Recommender System Using OpenAI Embedding Model

4.2 GPT and Lang Chain

This time, in order to address both research questions (R2 and R3), we propose an approach based on Retrieval Augmented Generation (RAG). This method, as seen in Fig. 2, is divided into two stages:

1. Preparation of our knowledge base.
  (i) Loading and Segmentation: The database constituting our analysis is loaded, and the new data is subsequently segmented into smaller and manageable units called chunks. The chunking process, commonly used in text or

data processing, is controlled by two main parameters. The chunk size determines the length of each piece, while the overlap between chunks indicates the amount of overlap between successive pieces

(ii) Embedding: These chunks are then transformed into a queryable format called the vector index.

(iii) Storage: The transformed data is stored in a vector database. For this second framework, we used another database, Chroma.

2. Course Recommendations:

(i) Definition of Prompts: In this phase, two distinct prompt models have been created. One is intended for general recommendations in case the user has no preferences, and the other prompt is dedicated to personalized recommendations to account for each user's specific preferences.

(ii) Configuration of RetrievalQA with Langchain and Use of ChatGPT 3.5 Turbo: We configure the RetrievalQA model with Langchain, leveraging the previously prepared knowledge base. Interaction with the user to obtain course-related queries is then facilitated by ChatGPT 3.5.

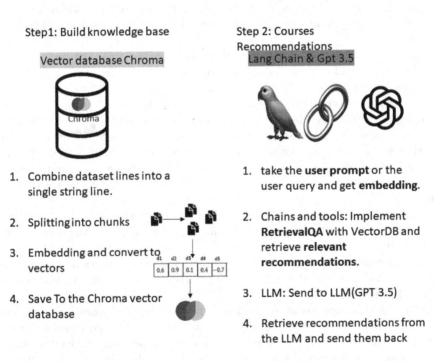

Fig. 2. Approach for Recommender System Using Gpt and Langchain

5 Results and Discussions

5.1 DataSet

To expand our database for our recommendation system, we combined resources from two online learning platforms. Data was extracted from the public websites of Coursera and Udacity in September 2021 and was manually entered in cases where the data extraction was done incorrectly. There are two types of datasets:

1. E-learning courses: This dataset contains over 622 courses with various characteristics, including the name, URL, description, and different course skills.
2. E-learning ratings: A list containing 1.45 million reviews with their ratings. This list also includes the name of the reviewer who wrote the review and the course identifier.

5.2 Embedding with Adda

To leverage OpenAI embedding models, it is imperative to register on the OpenAI platform and obtain an access key. Similarly, to use Pinecone as a vector database, creating an account and an index on the platform is essential to store the vectors.

As OpenAI is a paid platform, and in order to conserve API resources, we opted for the use of a reduced dataset consisting of 1000 users to build content-based recommendations. We begin by calculating the embedding vectors using the "text-embedding-ada-002" model, focusing on the "skills" column.

Using the Pinecone library, the normalized course vectors are indexed with their associated metadata, such as the course name. The user, through a request, can then specify the desired skills, and the system uses the "text-embedding-ada-002" model to calculate its embedding vector. This vector is then used to query the Pinecone index, which returns the top 10 most similar courses with their metadata.

The similarity measurement is based on cosine similarity and is calculated by comparing the embedding vector of the course with that of the user's query. A similarity score close to 1 indicates a strong match, suggesting that the course closely corresponds to the user's query. In Table 1, recommendations are provided with the similarity calculation for a user with a query related to the field of artificial intelligence.

Through this experiment, we addressed our first research question by demonstrating that the combination of OpenAI embedding with the Pinecone vector database represents a promising approach for a more personalized content-based recommendation. However, it is crucial to emphasize that the use of the OpenAI API may incur costs, requiring careful monitoring and management to avoid unexpected expenses.

At this stage, the use of a vector database becomes particularly relevant, allowing the storage of calculated embeddings for future reuse rather than recalculating them for each query. This approach provides an economical and efficient solution.

Table 1. E-learning Course Recommendations with Cosine Similarity

Id	Course Name	Similarity
86	Project Planning and Machine learning	0.822557807
19	Recommendation Systems with TensorFlow on GCP	0.822557807
57	Prediction and Control with Function Approximation	0.796458
65	Introduction to Recommender Systems: Non-Personalized and Content-Based	0.788251042
40	Real-time OCR and Text Detection with Tensorflow, OpenCV and Tesseract	0.785848439
91	Control physics with C in Unity	0.779547
58	Advanced Neurobiology I	0.778156

5.3 Langchain and GPT

Due to constraints in OpenAI's response time and cost, we made the decision to restrict our dataset to 1000 users. We randomly selected 10 courses per user to send to the chosen LLM model, which is GPT-3.5 Turbo in our case. To maintain balance in the test dataset, we extracted 5 courses with a rating higher than 3 and 5 courses with a rating lower than 3 for each user. We then combined all extractions to obtain a unique list in a single field.

The goal is to provide answers to our two research questions, R2 and R3. The following prompt template was adopted: "You are a content-based course recommender system that tailors recommendations based on the content of courses and user preferences. context" userinfo = "This is what we know about the user, and you can use this information to better tune your recommendations: "User Interests": + dftest['courseskills'][0] "Ratings given by the user": + dftest['coursesratings'][0] .

We use this prompt iteratively for each user in our dataset, querying our knowledge base constructed using the Chroma vector database this time. GPT generates a list of top recommended courses, as shown in Fig. 3.

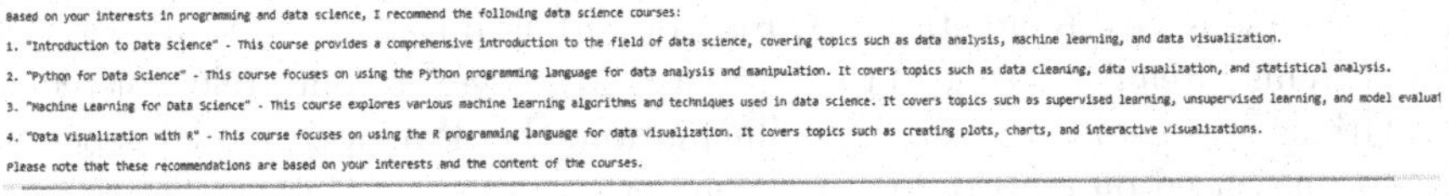

Based on your interests in programming and data science, I recommend the following data science courses:
1. "Introduction to Data Science" - This course provides a comprehensive introduction to the field of data science, covering topics such as data analysis, machine learning, and data visualization.
2. "Python for Data Science" - This course focuses on using the Python programming language for data analysis and manipulation. It covers topics such as data cleaning, data visualization, and statistical analysis.
3. "Machine Learning for Data Science" - This course explores various machine learning algorithms and techniques used in data science. It covers topics such as supervised learning, unsupervised learning, and model evaluation.
4. "Data Visualization with R" - This course focuses on using the R programming language for data visualization. It covers topics such as creating plots, charts, and interactive visualizations.
Please note that these recommendations are based on your interests and the content of the courses.

Fig. 3. Recommendations Generated Using GPT

However, using this prompt, we couldn't evaluate the accuracy of our framework. We obtained qualitative responses that didn't provide the predicted rating.

We modified the prompt, asking GPT to provide the predicted rating to assess its performance.

To calculate the performance of the proposed framework, we use two metrics: Hit Ratio and Recall@k. The Hit Ratio evaluates the success of recommendations by measuring the proportion of users for whom at least one recommendation led to an effective interaction. Recall@k, on the other hand, assesses the system's ability to recall relevant items among the top positions of the recommended list.

As a baseline, we used the Singular Value Decomposition (SVD) method for a collaborative recommendation system. Matrix factorization decomposes the user-item rating matrix into latent matrices, capturing the hidden features of users and courses. This model focuses on the similarity between users and courses based on their past ratings.

For k = 10 and k = 20, we obtained quite interesting results as mentioned in 2.

Table 2. GPT Recommendation System Metrics

Model	Recall10	Recall20	Hitratio
SVD	0,38	0,36	0,84
GPT 3.5 - Langchain	0,56	0,51	0,94

Moreover, to address our third research question, as can be seen in the results Table 2, GPT, through the use of Langchain and by utilizing the knowledge base constructed and stored in the Pinecone vector database, provides more effective results than the traditional recommendation system SVD chosen as the baseline.

Additionally, the prompts were designed so that when a user or a course is new (cold start), we use the retrieval model to obtain initial suggestions based on similar items already present in the database. Furthermore, to suggest relevant recommendations and avoid the issue of hallucination, the constructed model generates suggestions based on embeddings and semantic chains.

Furthermore, to answer our second research question, the results presented in Table 2 demonstrate that the use of GPT, combined with the Lang Chain method and leveraging a knowledge base stored in the Pinecone vector database, outperforms the traditional SVD recommendation system chosen as the reference. The formulation of prompts was designed to use the retrieval model when the user or course is new (cold start), thereby providing initial suggestions based on similar items already present in the database. Moreover, to offer relevant recommendations and avoid the issue of hallucination, the developed model generates suggestions based on embeddings and semantic chains.

6 Conclusion

Our work aimed to implement an embedding-based recommendation approach, leveraging the "Retrieval Augmented Generation" (RAG) model as the first approach and using Lang Chain in combination with RAG and GPT-3.5 as the second approach. The obtained results demonstrated that the use of GPT-3.5 generated significantly improved recalls compared to a collaborative SVD model used as a reference. Additionally, the RAG approach contributed to mitigating the limitations of natural language models such as GPT-3.5, taking into consideration issues of hallucination and challenges related to cold starts in recommendation systems. These results highlight the potential of embedding, Lang Chain, and the RAG approach in the context of recommendation systems, offering more efficient and personalized solutions. It is important to note that, despite promising performances, aspects such as the costs associated with the use of LLM models and the limitations of APIs must be taken into account, highlighting the need for careful management to avoid unforeseen expenses. Moreover, sustained attention must be paid to ethical concerns and the risks of bias that come with the use of these cutting-edge technologies. The tendency of LLM models to generalize from existing data raises questions about their ability to replicate or amplify undesirable biases in the recommendations made. This work paves the way for future research aimed at developing robust strategies to assess and minimize the impact of biases. Concurrently, it suggests the opportunity to deepen our research to explore new economic models to refine recommendation techniques based on embeddings and synergies between RAG, Lang Chain, and LLM models.

References

1. Lakshmi, G.J., Sresta, R.S.S., Kushmitha, C., Pavani, J.: A collaborative filtering based recommender system for hospital recommendation. In: 2023 7th International Conference on Trends in Electronics and Informatics (ICOEI), pp. 1309–1313, April 2023
2. RahmatAbadi, A.F., Mohammadzadeh, J.: Leveraging Deep Learning Techniques on Collaborative Filtering Recommender Systems. arXiv preprint arXiv:2304.09282 (2023)
3. Sinha, S., Sharma, T.: Content-based movie recommendation system: an enhanced approach to personalized movie recommendations. Int. J. Innov. Res. Comput. Sci. Technol. 11(3), 67–71 (2023)
4. Liu, F., Asaithambi, S.P.R., Venkatraman, R.: Hybrid personalized book recommender system based on big data framework. In: 2023 25th International Conference on Advanced Communication Technology (ICACT), pp. 333–340, February 2023
5. Le, N.L., Abel, M.H., Gouspillou, P.: Construction d'un systeme de recommandation base sur des contraintes via des graphes de connaissances arXiv preprint arXiv:2306.03247 (2023)
6. Fan, W., Zhao, Z., Li, J., Liu, Y., Mei, X., Wang, Y., Li, Q.: Recommender systems in the era of large language models (llms). arXiv preprint arXiv:2307.02046 (2023)
7. Lin, J., Dai, X., Xi, Y., Liu, W., Chen, B., Li, X., Zhang, W.: How can recommender systems benefit from large language models: a survey. arXiv preprint arXiv:2306.05817 (2023)

8. Fayyaz, Z., Ebrahimian, M., Nawara, D., Ibrahim, A., Kashef, R.: Recommendation systems: algorithms, challenges, metrics, and business opportunities. Appl. Sci. **10**(21), 7748 (2020)
9. Kamnis, S.: Generative pre-trained transformers (GPT) for surface engineering. Surface and Coatings Technology, 129680 (2023)
10. Worth, P.J.: Word embeddings and semantic spaces in natural language processing. Int. J. Intell. Sci. **13**(1), 1–21 (2023)
11. Siriwardhana, S., Weerasekera, R., Wen, E., Kaluarachchi, T., Rana, R., Nanayakkara, S.: Improving the domain adaptation of retrieval augmented generation (RAG) models for open domain question answering. Trans. Assoc. Comput. Linguist. **11**, 1–17 (2023)
12. Sachdev, A., Naik, A., Manhar, A.: Movie recommendation based system using time series data. Int. J. Sci. Res. Comput. Sci. Eng. Inf. Technol. (IJSRCSEIT), ISSN : 2456- 3307, Volume 9, Issue 3, pp. 455–458, May-June-2023
13. Labde, S., Karan, V., Shah, S., Krishnan, D.: Movie Recommendation System using RNN and Cognitive thinking. In: 2023 4th International Conference for Emerging Technology (INCET), pp. 1–7. IEEE, May 2023
14. Zhao, J., Su, B., Rao, X., Chen, Z.: A cross-platform personalized recommender system for connecting e-commerce and social network. Future Internet **15**(1), 13 (2022)
15. Sukestiyarno, Y.L., Sapolo, H.A., Sofyan, H.: Application of Recommendation System on E-Learning Platform Using Content-Based Filtering with Jaccard Similarity and Cosine Similarity Algorithms (2023)
16. Jena, K.K., Bhoi, S.K., Malik, T.K., Sahoo, K.S., Jhanjhi, N.Z., Bhatia, S., Amsaad, F.: E-learning course recommender system using collaborative filtering models. Electronics **12**(1), 157 (2022)
17. Hou, Y., Zhang, J., Lin, Z., Lu, H., Xie, R., McAuley, J., Zhao, W.X.: Large language models are zero-shot rankers for recommender systems. arXiv preprint arXiv:2305.08845 (2023)
18. Kang, W. C., Ni, J., Mehta, N., Sathiamoorthy, M., Hong, L., Chi, E., Cheng, D.Z.: Do LLMs Understand User Preferences? Evaluating LLMs On User Rating Prediction. arXiv preprint arXiv:2305.06474 (2023)

Data Management

Revolutionizing Skin Cancer Diagnosis: Unleashing AI Precision Through Deep Learning

Mohamad Abou Ali[1,2,3], Fadi Dornaika[1,4(✉)],
Ignacio Arganda-Carreras[1,4,5,6], Hussein Ali[2,3], and Malak Karaouni[2,3]

[1] University of the Basque Country (UPV/EHU), San Sebastian, Spain
{fadi.dornaika,ignacio.arganda}@ehu.eus
[2] Lebanese International University (LIU), Beirut, Lebanon
mohamad.abouali01@liu.edu.lb
[3] Beirut International University (LIU), Beirut, Lebanon
[4] Ikerbasque, Basque Foundation for Science, Bilbao, Spain
[5] Donostia International Physics Center (DIPC), San Sebastian, Spain
[6] Biofisika Institute (CSIC, UPV/EHU), Leioa, Spain

Abstract. Accurate detection methods are desperately needed, as skin cancer concerns around the world continue to rise. Conventional methods, which depend on dermatologists' subjective visual evaluation, face difficulties that highlight the critical role of artificial intelligence (AI). This study leverages the ISIC2019 dataset and advanced deep learning models to tackle the challenging job of 8-class skin cancer classification. Multiple pre-trained Vision Transformer models and ImageNet topologies are used in an extensive examination. An unique "Naturalize" augmentation technique is presented to resolve intrinsic class imbalances, which result in the construction of the pioneering Naturalized 7.2K ISIC2019 and the groundbreaking 2.4K ISIC2019 datasets, improving classification accuracy. The study emphasizes how important AI is in reducing the possibility of an incorrect or underdiagnosed case of skin cancer. The ability of AI to analyze large datasets and identify subtle trends greatly enhances diagnostic skills. Early detection and better patient outcomes are promised by the integration of AI-powered technologies. Within the Naturalized 7.2K ISIC2019 dataset, quantitative measurements such as confusion matrices and visual assessments utilizing Score-CAM demonstrate an unparalleled success: 100% accuracy, precision, recall, and F1-Score. In conclusion, this study shows how AI-powered systems can significantly improve skin cancer diagnosis, marking a noteworthy development in the area of dermatological diagnostics. The Naturalized 7.2K ISIC2019 dataset demonstrated 100% performance across crucial parameters, demonstrating the revolutionary power of AI-driven techniques that are changing the face of skin cancer detection and patient treatment.

Keywords: Skin Cancer · Convolutional Neural Net (CNN) · Data augmentation · Naturalize · Pre-Trained Models · Deep Learning (DP)

1 Introduction

Millions of people worldwide are impacted by skin cancer, which can have fatal implications in addition to significant medical expenses and physical disfigurement. An alarming 9,500 new cases of skin cancer are reported every day [17], suggesting that a fifth of Americans may get the disease. In addition to the physical cost, invasive procedures and obvious scars from treatments frequently result in emotional discomfort.

Excessive ultraviolet (UV) radiation exposure, whether from the solar or through artificial means like tanning beds [8], is what causes skin cancer in the first place. Those with fair skin, light eyes, and hair are more vulnerable because they have less melanin, the body's natural UV shield. A history of sunburns, a genetic susceptibility, and specific chemical exposures are risk factors. The ISIC 2019 dataset [5] provides dermoscopic images for computer-aided diagnosis, hence supporting research on skin cancer.

In 2019, the ISIC 2019 dataset [5] significantly influences machine learning, computer vision, and AI for early skin cancer detection. Researchers use it to train, validate, and test algorithms, particularly in deep learning, transforming skin cancer detection and improving diagnostic tools for healthcare professionals.

This research investigates the use of deep learning methods for the categorization of skin cancer, incorporating models like ImageNet pretrained CNNs [18] and Google ViT model [7] by the general use of the transfer learning (DL) technique and its particular kind, fine-tuning. Assessment methods include visual inspections with tools such as Channel Attention Module (CAM), e.g. Score-CAM [21], and quantitative assessments with confusion matrices and related classification reports. Our methodology integrates advanced deep learning models and the "Naturalize" technique [1], representing a significant advancement in automated skin cancer classification.

There are five portions to this work: A thorough review of pertinent literature will be included in Sect. 2; the study's methodology will be covered in Sect. 3; the experimental findings utilizing Google ViT and ImageNet CNNs will be shown in Sect. 4; and a thorough conclusion will be provided in Sect. 5.

2 Previous Works

Deep learning models for skin lesion categorization have advanced significantly in the last several years. This section summarizes findings from significant research that used several CNN architectures to address this problem, with an emphasis on approaches and results with the ISIC2019 dataset.

Using the GoogleNet (Inception V1) model utilized transfer learning, Kassem et al. [11] achieved an astounding 94.92% accuracy on the ISIC2019 dataset. In terms of precision (80.36%), recall (79.80%), and F1-score (80.07%), their model performed remarkably well. The Quantum Inception-ResNet-V1 was introduced by Li et al. [14] in 2022, and using the ISIC2019 dataset, it achieved an astounding 98.76% accuracy. The model demonstrated notable enhancements in F1-score (98.33%), recall (98.26%), and precision (98.40%), indicating a noteworthy increase in accuracy. By combining MobileNet with transfer learning, Mane et al. [16] were able to achieve an accuracy of 83% on the ISIC2019 dataset. Even with relatively poorer performance, their model demonstrated stable recall, precision, and F1-score at 83%, indicating strong classification. With an accuracy of 84.80%, Hoang et al. [9] presented the Wide-ShuffleNet in combination with segmentation approaches. In contrast to earlier research, their proposed architecture displayed reduced metrics for F1-score (72.61%), recall (70.71%), and precision (75.15%). IA four-layer DCNN model was introduced by Fofanah et al. [3] in 2023, and it achieved 84.80% accuracy on a modified dataset split. In terms of recall (83.80%), precision (80.50%), and F1-score (81.60%), their model performed in balance. A Residual Deep CNN model was also proposed by Alsahaf et al. [2] in the same year, and it achieved 94.65% accuracy on a different dataset split. Recall (70.78%), precision (72.56%), and F1-score (71.33%) were all measured in a balanced manner. Taken as a whole, these research show how skin lesion categorization algorithms have developed, with notable advancements in accuracy and performance measures. The merits and demerits of each model are emphasized through a comparison study, which establishes the groundwork for future developments in dermatological image categorization.

A brief summary of these efforts is given by the literature review, which focuses on a number of publications (Table 1) that are concerned with automating skin cancer categorization using the ISIC2019 dataset.

Table 1. An summary of relevant literature

Citation	Dataset	Data Split	Accuracy	Recall	Precision	F1-Score
[11]	ISIC2019	80/10/10	94.92	79.80	80.36	80.07
[14]	ISIC2019	80/10/10	98.76	98.26	98.40	98.33
[16]	ISIC2019	80/10/10	83	83	83	82
[9]	ISIC2019	90/10	84.80	70.71	75.15	72.61
[3]	ISIC2019	60/10/30	84.80	83.80	80.50	81.60
[2]	ISIC2019	70/15/15	94.65	70.78	72.56	71.33

Class imbalance and data scarcity are addressed in deep learning using our novel 'Naturalize' augmentation technique. Our skin cancer classification model's exceptional testing accuracy, precision, recall, and F1-score averaged at 100% were achieved by applying 'Naturalize'. In addition to improving classification performance, this novel approach revolutionizes deep learning by changing the face of accurate diagnosis in disparate and heterogeneous skin cancer categories.

3 Materials and Methods

This part describes our state-of-the-art method for accurately classifying skin cancer images on the difficult ISIC2019 dataset. Modern Deep Learning techniques were used, most notably "ImageNet pre-trained CNNs" and "Google Vision Transformers (ViT)." To address data insufficiencies in specific classes (AK, DER, VAS, SCC) and boost performance, we integrated the innovative "Naturalize" augmentation technique. We equalized image numbers across classes by reducing three classes to 2.4k images, aligning them with the BKL class, aiming to eliminate imbalances among MEL, NV, BCC classes (ranging from 10.4k to 1k images). Figure 1 visually showcases the breadth and sophistication of our methodology, highlighting our dedication to advancing image classification.

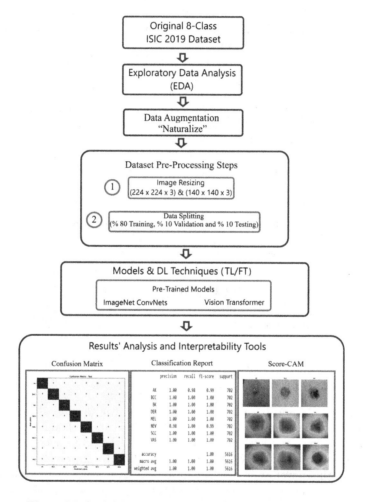

Fig. 1. Methodology workflow using the ISIC 2019 dataset.

3.1 ISIC-2019 Dataset

Published ISIC 2019 Dataset There are many different types of skin cancer, but for the purposes of this study, we will focus on eight classes: Actinic Keratosis (AK), Basal Cell Carcinoma (BCC), Benign Keratosis (BK), Melanocytic Nevi (NEV), Melanoma (MEL), Vascular Skin Lesion (VAS), Dermatofibroma (DER), and Squamous Cell Carcinoma (SCC). The published ISIC 2019 dataset [5] was gathered from an internet archive and comprises 25,331 photos divided into these eight groups that correspond to various forms of skin cancer.

We changed the ISIC-2019 dataset to correct the uneven distribution of photos. Specifically, we reduced the count of images for three different classes of skin cancer (MEL, NV, and BCC) to 2.4k and aligned them with the current count of 2.4k images corresponding to the BK class. To create equilibrium among the many forms of cancer, this change was done. For this, we used the Naturalize Augmentation approach. As a result, 19,200 balanced photos representing the eight different forms of skin cancer are included in the revised dataset.

A summary of how the eight kinds of skin cancer are distributed in the ISIC 2019 dataset can be seen in Table 2 [2].

Table 2. ISIC-2019 dataset - Classes' representation.

#	Skin cancer class	Images' count	(%)
1	AK	867	3.322
2	BCC	3,323	13.11
3	Bk	2,624	10.35
4	DER	239	0.94
5	NEV	12,875	50.82
6	MEL	4,522	17.85
7	VAS	253	1.138
8	SCC	628	2.47
	Total	25,331	100

The ISIC dataset's photos are 1024 × 1024 pixels in size by default [2]; however, to increase flexibility in use, they must be adjusted to "224 × 224" and "140 × 140".

Naturalized ISIC2019 Datasets
The objective was to attain an equivalent quantity of images for each among the eight categories of skin cancer. To this end, the Naturalize enhancement is utilized. Using the Naturalize augmentation technique, two revised, balanced versions of ISIC2019 are produced: ISIC2019 datasets that have been naturalized: 2.4K and 7.2K.

Table 3 summarizes the distribution of the Naturalized 2.4K ISIC2019 dataset in the 8 classes.

Table 3. Naturalized 2.4K ISIC-2019 dataset - Classes' representation.

#	Skin cancer class	Images' count	(%)
1	AK	2,400	12.5
2	BCC	2,400	12.5
3	BK	2,400	12.5
4	DER	2,400	12.5
5	NEV	2,400	12.5
6	MEL	2,400	12.5
7	VAS	2,400	12.5
8	SCC	2,400	12.5
	Total	19,200	100

The distribution of the Naturalized 7.2K ISIC2019 dataset across the eight classes is summarized in Table 4.

Table 4. Naturalized 7.2K ISIC-2019 dataset - Classes' representation.

#	Skin cancer class	Images's count	(%)
1	AK	7,200	12.5
2	BCC	7,200	12.5
3	BK	7,200	12.5
4	DER	7,200	12.5
5	NEV	7,200	12.5
6	MEL	7,200	12.5
7	VAS	7,200	12.5
8	SCC	7,200	12.5
	Total	57,600	100

With distinctive quality and details that mimic the original ISIC2019 dataset, the Naturalize method can produce an infinite number of skin cancer photos. This is made possible by adding segmented skin cancer photos with various skin tones, which harness the power of unpredictability to their advantage.

3.2 Exploratory Data Analysis (EDA)

Analyzing exploratory data (EDA) using the ISIC 2019 dataset revealed significant class imbalances, particularly in DER and NEV categories, impacting overall model performance metrics. To address this, the "Naturalize" augmentation technique was employed, generating quality images for underrepresented classes.

This effectively mitigated imbalances and maintained image quality, enhancing overall model performance.

3.3 Data Augmentation "Naturalize"

The "Naturalize" augmentation technique's working concept is illustrated in the pseudocode presented in Algorithm 1. Naturalize technique is comprised of two main stages:

First-Segmentation 2: Using the "Segment Anything Model (SAM)" [13], photos representing four distinct forms of skin cancer were separated into smaller subsets within the ISIC 2019 dataset. Segments for AK, DER, VAS, and SCC were developed by this technique. According on the classification report and performance indicators from the previous EDA analysis, adding these new photos to the classes improved the classification accuracy.

Step 2—Producing Composite Pictures (Fig. 3)
We combined randomly selected photos of normal skin with the four SAM-segmented categories (AK, DER, VAS, and SCC) to create new composite images. Using the production of composite skin cancer images as an example, this technique is visually depicted in Figs. 2 and 3.

Algorithm 1. The Steps in Naturalize Pseudocode Algorithm

1: # **Paths & Libraries**
2: Import os, random, image_processing, SAM_model
3: Set up file paths and import necessary libraries.
4:
5: # **Loading ISIC 2019 Dataset & SAM_model**
6: Mounting Google_drive
7: Loading ISIC 2019 dataset's Skin Cancer photos
8: SAM = load_model(SAM_model)
9:
10: # **Using SAM to segment ISIC 2019 Dataset**
11: Using SAM to segment ISIC 2019 images into segmented "Cancer" images
12: Saving "Cancer" images into new "Skin Cancer" dataset created on Google_drive
13:
14: # **The random choosing of Skin Background dataset**
15: Selecting at random Skin Background image from Skin Background dataset
16:
17: # **Creating Composite Images**
18: **for** i in range(num_images) **do**
19:     Loading Skin Background image
20:     Choosing arbitrarily "Cancer" image from "Skin Cancer" dataset
21:     Rotating at random "Cancer" image
22:     Adding "Cancer" image at random position to Skin Background image
23:     Saving the created composite image on Google_drive
24: **end for**

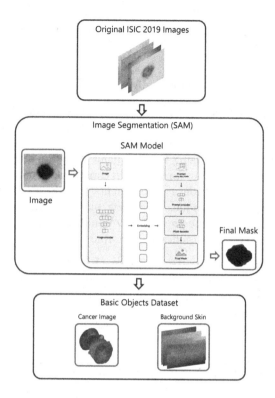

Fig. 2. The "Naturalize" first step—segmentation.

3.4 Comparing Conventional Augmentation Techniques with Naturalize

"Naturalize" method surpasses traditional image augmentation techniques by utilizing a deliberate segmentation procedure with the "Segment Anything Model." This approach isolates certain classes of skin cancer, leading to a substantial increase in segmented instances. The integration of these segmented objects results in a diverse dataset, offering precision control over image quality through random rotation and placement within background skin images. This method, initially developed for medical imaging, exhibits versatility across domains, highlighting its potential for widespread application beyond the medical field.

3.5 Preparation of the Naturalized 2.4K and 7.2K ISIC2019 Datasets

Two main processes were engaged in the preparation of ISIC2019 Naturalized 2.4K and 7.2K datasets:

Step 1-Resizing Images: The photos were downsized to fit the usual input size of "224 × 224" used for pre-trained ViT and ImageNet ConvNets models. In

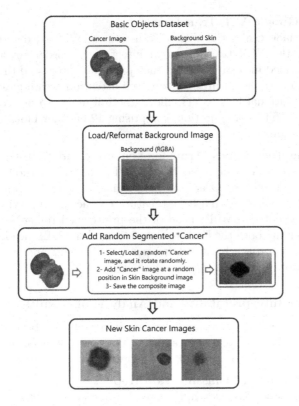

Fig. 3. The "Naturalize" second step—composite image generation.

an effort to maximize computational resources, the photos were also downsized to dimensions of "140 × 140," particularly with a large dataset such as the Naturalized 7.2K dataset.

Step 2-Data Separation: The Naturalized 2.4K and 7.2K datasets were split into three subsets: a training set consisting of 80%, a validation set of 10%, and a testing set of 10%.

3.6 Deep Learning Models and Techniques

This study used two different types of model architectures: pre-trained Google Vit and pre-trained ImageNet CNNS. Furthermore, two DL methods [23] were used to train pre-trained models: Fine-Tuning (FT) and Transfer Learning (TL).

Pre-trained ImageNet CNNs
ConvNets are explicitly demonstrated using pre-trained ImageNet models, which undergo extensive training using a substantial dataset. The essential component of this work was the use of pre-trained ImageNet models. Among the notable models used in this study were DenseNet-201 [10], EfficientNetV2 B0 [20], InceptionResNet [19], Xception [4], VGG16 [15], and VGG-19 [15].

Pre-trained Google ViT Transformer
The research made use of the Vision Transformer (ViT) [7] design, which was modified from the NLP transformer architecture. In order to do this, input images were divided into patches, and each patch was processed through Google ViT encoder to extract features utilizing self-attention techniques. Unlike conventional convolutional layers, ViT allows the full image to be analyzed simultaneously. The "ViT" configuration, comprising 12 encoder blocks, was utilized in the investigation.

Deep Leaning Techniques: Transfer Learning and Fine-tuning
Transfer Learning (TL) and Fine-Tuning (FT) are two methods used in this work to create an efficient Deep Learning skin cancer tool. Transfer Learning entails fixing the Convolutional Base (Feature Extractor) but switching out the Multi-Layer Perceptron (MLP) head of the pre-trained ImageNet model for a new job. By training both the Feature Extractor and the MLP head, Fine-Tuning expands on this approach by modifying parameters to enhance performance on a particular learning task [12,22].

3.7 Tools for Interpretability and Analysis of Results

Three tools are used for result analysis and interpretation beside to the accuracy measures, which include accuracy and loss. The confusion matrix with its corresponding classification reports, and Score-CAM are the tools in this set.

Confusion Matrix and Classification Report
Particularly in supervised learning contexts, an error matrix, or confusion matrix [6], offers a visual picture of an algorithm's performance. Real classes are shown in the rows, and classes that are anticipated are shown in the columns. Metrics like precision, recall, and F1-score for each class are used in the assessment to evaluate the quality of the predictions. It also has weighted average and macro accuracy to assess overall performance.

Score-CAM
According to reference [21], a Score-CAM is a technique for visual explanation that uses CNN models' Class Activation Mapping (CAM) to apply weights to scores. It accomplishes the goal of illuminating internal mechanisms of CNN models.

4 Results

The studies on eight-class skin cancer classification with the ISIC2019 dataset using a variety of models-including pre-trained ImageNet and ViT models-are summarized in this section. The "Naturalize" augmentation method produced two balanced datasets in order to address class imbalance. Using published ISIC2019 dataset, and the new Naturalized 2.4K Naturalized 7.2K datasets, Score-CAM was used to visually analyze model performance. Confusion matrices and classification reports were used to statistically compare the performance of models.

4.1 Findings from the Naturalized 2.4K Dataset

Initially, transfer learning (TL) was applied to all pre-trained models; however, fine-tuning (FT) yielded superior results. Accuracy ratings for the Naturalized 2.4K dataset's training, validation, and testing subsets are shown in Table 5. ConvNexTBase obtained the best training accuracy, whereas DenseNet201 demonstrated superior validation and overall accuracy.

Table 5. Naturalized 2.4K dataset -An overview of the training, validation, and testing accuracy of the models.

Experimental Models	Average Accuracy		
	Train (%)	Validate (%)	Test (%)
ConvNexTBase	99	95	92
ConvNeXtLarge	87	84	84
DenseNet-201	97	95	95
EfficientNetV2 B0	88	85	82
InceptionResNetV2	94	90	89
VGG16	97	93	94
VGG-19	96	89	90
ViT	89	87	90
Xception	94	91	82

The F1-score, recall and precision are displayed in Table 6 for the Naturalized 2.4K testing dataset. Notable is the DenseNet-201 model at maximum performance.

After a series of trials, the DenseNet-201 model was picked because of its superior performance in the testing and validation subsets.

4.2 The Findings of DenseNet-201 Model

Table 7 displays the confusion matrix classification report of the FT DenseNet-201 model using the publicly available ISIC 2019 dataset.

The classification report of the FT DenseNet201 model using the Naturalized 2.4k dataset is presented in Table 8.

Using the Naturalized 7.2K dataset, the classification report of the FT DenseNet201 model is shown in Table 9.

5 Discussion

An extensive examination of the experiments for an ambitious eight-class skin cancer classification problem is provided in the Discussion section. Using the

Table 6. Naturalized 2.4K dataset - An overview of the macro-average precisions, recalls, and F1-scores of the models.

Experimental Models	Macro Average		
	Precision (%)	Recall (%)	F1-Score (%)
ConvNexTBase	93	92	91
ConvNeXtLarge	87	86	87
DenseNet-201	96	95	95
EfficientNetV2 B0	86	82	80
InceptionResNetV2	90	89	88
VGG16	94	94	94
VGG-19	90	90	89
ViT	91	90	90
Xception	86	87	86

Table 7. ISIC2019 dataset—Classification report of FT DenseNet-201 Model.

Skin Cancer	Precision (%)	Recall (%)	F1-Score (%)	Support
AK	61	67	60	66
BCC	74	69	79	333
BK	58	88	79	263
DER	56	75	69	24
NEV	88	93	90	1287
MEL	65	36	46	452
VAS	85	87	89	63
SCC	75	94	86	25
Accuracy			78	2,513
Macro Average	76	68	70	2,513
Weighted Average	85	81	81	2,513

ISIC2019 dataset, a number of models-including Vision Transformer and pre-trained ImageNet architectures-were carefully evaluated. Two balanced datasets were produced as a result of the novel "Naturalize" augmentation technique, which was used to correct class imbalance. The quantitative evaluation involved the use of confusion matrices and classification reports; it was further improved by a visual analysis procedure utilizing Score-CAM on the ISIC2019, Naturalized 2.4K, and 7.2K datasets.

5.1 Naturalized 2.4k Dataset - Findings' Analysis

Transfer learning (TL) was first used on pre-trained CNNs and ViT models, but fine-tuning (FT) resulted in a notable improvement. The Naturalized 2.4K

Table 8. Naturalized 2.4K dataset—Classification report of FT DenseNet-201 Model.

Skin Cancer	Precision (%)	Recall (%)	F1-Score (%)	Support
AK	98	98	98	240
BCC	99	98	99	240
BK	95	100	97	240
DER	100	100	100	240
NEV	85	98	91	240
MEL	99	80	89	240
VAS	100	100	100	240
SCC	100	99	99	240
Accuracy			97	1920
Macro Average	97	97	97	1920
Weighted Average	97	97	97	1920

Table 9. Naturalized 7.2K dataset—Classification report of FT DenseNet-201 Model.

Skin Cancer	Precision (%)	Recall (%)	F1-Score (%)	Support
AK	100	98	99	760
BCC	100	100	100	760
BK	100	100	100	760
DER	100	100	100	760
NEV	98	100	99	760
MEL	100	100	100	760
VAS	100	100	100	760
SCC	100	100	100	760
Accuracy			100	5760
Macro Average	100	100	100	5760
Weighted Average	100	100	100	5760

dataset accuracy scores are displayed in Table 6, where DenseNet201 performed best in testing and validation while ConvNexTBase lead in training accuracy. DenseNet-201's dominance on the testing subset was further validated by the F1-score, recall and precision presented in Table 7. Owing to its exceptional performance, FT DenseNet-201 model was chosen for further experiments.

5.2 FT DenseNet-201 - Findings' Analysis

The confusion matrices classification reports of the FT DenseNet-201 model on the published ISIC2019, Naturalized 2.4K, and Naturalized 7.2K I datasets are shown in Tables 7, 8, and 9.

The capacity of "Naturalize" to produce high-quality photos broadens its application outside the medical field, mitigating class disparities and improving model performance in a variety of classification tasks. The FT DenseNet-201 model's performance on the ISIC2019 datasets is summarized in Table 10, which shows notable gains in accuracy, F1-score, recall, and macro-average precision thanks to balanced datasets. The Naturalized 7.2K dataset, in particular, displays remarkable results. DenseNet-201 receives flawless scores across the board, highlighting the effectiveness of the "Naturalize" strategy in improving the accuracy of skin cancer classification.

Table 10. ISIC2019 datasets (Published, Naturalized 2.4K and 7.2K)—Classification reports' summaries of FT DenseNet-201 model.

PBC Datasets	Macro Average			
	Precision (%)	Recall (%)	F1-Score (%)	Accuracy (%)
Imbalanced ISIC2019 Datasets				
Published	76	68	70	93
Naturalized Balanced ISIC2019 Datasets				
Naturalized 2.4K	97	97	97	97
Naturalized 7.2K	100	100	100	100

5.3 Score-CAM Interpretability

The results of the investigation were expanded and reinforced by the use of Score-CAM, a method of interpretability. The FT DenseNet201 model's performance on the Naturalized 7.2K dataset is shown graphically in Fig. 4. This visualization improves our comprehension of skin cancer categorization by verifying accurate classifications and openly highlighting important picture regions in the decision-making process. In addition to validating model performance, Score-CAM provides insightful information about critical regions of the image that need to be classified.

5.4 Work's Novelty

The evaluation against prior studies is divided into two segments. In the initial phase of comparison, the testing dataset exclusively originates from the published ISIC2019 dataset, ensuring a consistent foundation for benchmarking against earlier works. This alignment is succinctly presented in Table 11.

The performance metrics of several previous research are compared with our approach in Table 11 for the ISIC2019 dataset skin cancer classification. Previous studies show a range of recalls, F1-Scores, accuracies, and precisions, which

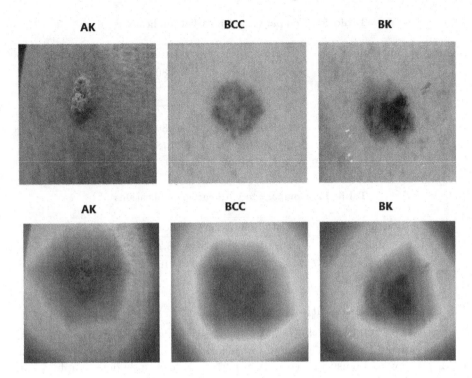

Fig. 4. Score-CAM for fine-tuned DenseNet-201.

demonstrate different results. Among all the methods, ours achieves an exceptional 97% score for accuracy, F1-Score, precision, and recall. This impressive findings represents a revolutionary breakthrough in the categorization of skin cancer and highlights the efficiency and dependability of our approach in comparison to other current approaches.

In the second phase of the comparison, the original ISIC 2019 dataset undergoes augmentation using the Naturalize augmentation technique. The resultant enhanced dataset, called "Naturalized 7.2K," is then split into three parts: 80% train, 10% validate, and 10% test. The performance assessment is conducted based on this division, following a methodology akin to the one employed in the previous work [14], as outlined in Table 12.

In Table 11, we use an identical number of testing images as the original ISIC 2019 dataset, excluding any taken from the Naturalize method. Conversely, in Table 12, the regenerated dataset, referred to as "Naturalized 7.2K," undergo a new allocation for Training, Validation, and Testing.

Performance metrics for skin cancer classification (ISIC2019 dataset) are compared between our method and previous work [14] in Table 12. Our method achieves perfect scores (100%) regarding F1-Score, recall, precision, and accuracy, indicating a noteworthy breakthrough in this area.

Table 11. Comparison with earlier publications

Citation	Dataset	Data Split	Accuracy	Recall	Precision	F1-Score
[11]	ISIC2019	80/10/10	94.92	79.80	80.36	80.07
[16]	ISIC2019	80/10/10	83	83	83	82
[9]	ISIC2019	90/10	84.80	70.71	75.15	72.61
[3]	ISIC2019	60/10/30	84.80	83.80	80.50	81.60
[2]	ISIC2019	70/15/15	94.65	70.78	72.56	71.33
Ours	Naturalized 2.4K	80/10/10	97	97	97	97

Table 12. Comparison with earlier publications

Citation	Dataset	Data Split	Accuracy	Recall	Precision	F1-Score
[14]	ISIC2019	80/10/10	98.76	98.26	98.40	98.33
Ours	Naturalized 7.2K	80/10/10	100	100	100	100

5.5 Ethical Considerations

Our study on AI-driven skin cancer diagnosis underscores ethical concerns, notably surrounding data privacy and algorithmic bias. Safeguarding patient data privacy and ensuring equitable performance across demographics are paramount. Future research will prioritize examining these ethical dimensions to promote responsible and fair deployment of AI diagnostic tools in healthcare.

5.6 Discussion Summary

"Naturalize" demonstrates unparalleled efficacy in generating high-quality, diverse images for skin cancer diagnosis. Its capacity to produce authentic replicas eliminates the need for additional augmentation techniques, surpassing traditional limitations. By autonomously generating precise images, "Naturalize" emerges as the premier augmentation tool, rendering alternative methods obsolete.

6 Conclusions

The goal of this project is to improve accessibility and precision in the diagnosis of skin cancer by utilizing artificial intelligence. Using the published ISIC2019 dataset and advanced DL models, the study investigates the eight-class classification of skin cancer and reveals the subtleties of diagnosis. By correcting for unequal distribution of classes, the "Naturalize" method produces meaningful datasets with previously unheard-of classification accuracy. The Naturalized 7.2K dataset's attainment of 100% metrics highlights AI's potential to be a game-changer in reducing the risk of misdiagnosis. With an emphasis on the pressing need to incorporate AI into clinical practice for better patient care, the study represents a paradigm change in dermatological diagnosis.

References

1. Abou Ali, M., Dornaika, F., Arganda-Carreras, I.: Blood cell revolution: Unveiling 11 distinct types with naturalize augmentation. Algorithms **16**(12) (2023). https://doi.org/10.3390/a16120562, https://www.mdpi.com/1999-4893/16/12/562
2. Alsahafi, Y., Kassem, M., Hosny, K.: Skin-net: a novel deep residual network for skin lesions classification using multilevel feature extraction and cross-channel correlation with detection of outlier. J. Big Data **10**, 105 (2023). https://doi.org/10.1186/s40537-023-00769-6
3. Balla Fofanah, A., Özbtlge, E., Kirsal, Y.: Skin cancer recognition using compact deep convolutional neural network. Çukurova Üniversitesi Mühendislik Fakültesi Dergisi **38**(3), 787-797 (2023).https://doi.org/10.21605/cukurovaumfd.1377752
4. Chollet, F.: Xception: Deep learning with depthwise separable convolutions. CoRR **abs/1610.02357** (2016). http://arxiv.org/abs/1610.02357
5. Codella, N., et al.: Skin lesion analysis toward melanoma detection 2018: A challenge hosted by the international skin imaging collaboration (isic) (2019)
6. Dalianis, H.: Evaluation Metrics and Evaluation, pp. 45–53. Springer International Publishing, Cham (2018). https://doi.org/10.1007/978-3-319-78503-5_6
7. Dosovitskiy, A., et al.: An image is worth 16x16 words: Transformers for image recognition at scale. CoRR **abs/2010.11929** (2020). https://arxiv.org/abs/2010.11929
8. Garrubba, C., Donkers, K.: Quick recertification series jaapa skin cancer. JAAPA: official journal of the American Academy of Physician Assistants (01 2020). https://doi.org/10.1097/01.JAA.0000651756.15106.3e
9. Hoang, L., Lee, S.H., Lee, E.J., Kwon, K.R.: Multiclass skin lesion classification using a novel lightweight deep learning framework for smart healthcare. Appl. Sci. **12**(5) (2022). https://doi.org/10.3390/app12052677, https://www.mdpi.com/2076-3417/12/5/2677
10. Huang, G., Liu, Z., Van Der Maaten, L., Weinberger, K.Q.: Densely connected convolutional networks. In: 2017 IEEE Conference on Computer Vision and Pattern Recognition (CVPR), pp. 2261–2269 (2017). https://doi.org/10.1109/CVPR.2017.243
11. Kassem, M.A., Hosny, K.M., Fouad, M.M.: Skin lesions classification into eight classes for isic 2019 using deep convolutional neural network and transfer learning. IEEE Access **8**, 114822–114832 (2020). https://doi.org/10.1109/ACCESS.2020.3003890
12. Kim, H., Cosa-Linan, A., Santhanam, N., Jannesari, M., Maros, M., Ganslandt, T.: Transfer learning for medical image classification: a literature review. BMC Med. Imaging **22** (2022). https://api.semanticscholar.org/CorpusID:248132791
13. Kirillov, A., et al.: Segment anything (2023)
14. Li, Z.H., et al.: A classification method for multi-class skin damage images combining quantum computing and inception-resnet-v1. In: Frontiers of Physics (2022). https://api.semanticscholar.org/CorpusID:253260727
15. Liu, S., Deng, W.: Very deep convolutional neural network based image classification using small training sample size. In: 2015 3rd IAPR Asian Conference on Pattern Recognition (ACPR), pp. 730–734 (2015).https://doi.org/10.1109/ACPR.2015.7486599
16. Mane, D., Ashtagi, R., Kumbharkar, P., Kadam, S., Salunke, D., Upadhye, G.: An improved transfer learning approach for classification of types of cancer. Traitement du Signal **39**, 2095–2101 (2022). https://doi.org/10.18280/ts.390622

17. Siegel, R., Miller, K., Fuchs, H., Jemal, A.: Cancer statistics, 2022. CA: A Cancer Journal for Clinicians **72** (01 2022). https://doi.org/10.3322/caac.21708
18. Stock, P., Cisse, M.: Convnets and imagenet beyond accuracy: understanding mistakes and uncovering biases. In: Ferrari, V., Hebert, M., Sminchisescu, C., Weiss, Y. (eds.) Computer Vision - ECCV 2018, pp. 504–519. Springer, Cham (2018)
19. Szegedy, C., Ioffe, S., Vanhoucke, V.: Inception-v4, inception-resnet and the impact of residual connections on learning. CoRR abs/1602.07261 (2016). http://arxiv.org/abs/1602.07261
20. Tan, M., Le, Q.: Efficientnetv2: Smaller models and faster training. CoRR abs/2104.00298 (2021). https://arxiv.org/abs/2104.00298
21. Wang, H., Wang, Z., Du, M., Yang, F., Zhang, Z., Ding, S., Mardziel, P., Hu, X.: Score-cam: score-weighted visual explanations for convolutional neural networks. In: 2020 IEEE/CVF Conference on Computer Vision and Pattern Recognition Workshops (CVPRW), pp. 111–119 (2020). https://doi.org/10.1109/CVPRW50498.2020.00020
22. Yin, X., Chen, W., Wu, X., Yue, H.: Fine-tuning and visualization of convolutional neural networks. In: 2017 12th IEEE Conference on Industrial Electronics and Applications (ICIEA), pp. 1310–1315 (2017). https://doi.org/10.1109/ICIEA.2017.8283041
23. Zhang, A., Lipton, Z., Li, M., Smola, A.: Dive into deep learning. CoRR abs/2106.11342 (2021). https://arxiv.org/abs/2106.11342

Fuzzy Bayesian Network Applied to Modeling Vehicles Cooling Systems Failure Risk

Soulaimane Idiri(✉), Hafida Khalfaoui, and Abdellah Azmani

Intelligent Automation and BioMedGenomics Laboratory, Faculty of Sciences and Technique of Tangier, Abdelmalek ESSAADI University Tetouan, Tetouan, Morocco
soulaimane.idiri@etu.uae.ac.ma, a.azmani@uae.ac.ma

Abstract. In today's competitive industrial landscape, the logistics sector faces escalating challenges in meeting customer demands while maintaining operational efficiency. One critical aspect is the risk of failure in vehicles cooling systems, which can disrupt production and distribution processes. To address this, risk assessment methodologies are crucial. In this study, we propose a novel approach utilizing Fuzzy Bayesian Networks (FBN) to assess the risk of vehicle's cooling system failures. Leveraging Fuzzy Logic and experts' knowledge, the FBN model integrates uncertain human knowledge to generate rules and calculate Conditional Probabilities (CPs). This method enables a comprehensive evaluation of the impact of various parameters on cooling system reliability. By amalgamating the causal relationship graph with CPs derived from the Fuzzy Logic system, our approach provides insights into the likelihood of system failures across different scenarios. Through the application of advanced computational tools such as FisPro and OpenMarkov, we validate our model and analyze the sensitivity of key factors affecting system performance. Our findings offer valuable insights for proactive maintenance strategies aimed at minimizing downtime and optimizing system reliability in industrial settings.

Keywords: Risk assessment · Vehicles cooling systems failure risk · Fuzzy Bayesian Network

1 Introduction

In the highly competitive logistics industry, characterized by its relentless pursuit of enhanced performance metrics, the exponential growth in the industrial sector has prompted heightened customer expectations [1]. Consequently, enterprises face the imperative to adopt proactive measures to mitigate disruptions in production and distribution value chains. Despite technological advancements, many companies grapple with reactive maintenance practices, resulting in significant costs and productivity losses [2].

To address these challenges, the concept of predictive maintenance has emerged as a transformative strategy [3]. Unlike traditional reactive or preventive approaches, predictive maintenance empowers maintenance teams to preemptively anticipate potential breakdowns. By optimizing resource allocation and minimizing downtime, predictive maintenance offers substantial cost savings and enhances operational efficiency [4, 5].

In this context, our study endeavors to unravel the intricate interplay of factors influencing vehicles cooling systems. Leveraging a sophisticated approach, we employ Bayesian networks to construct a robust model capable of illuminating critical parameters impacting system's performance. This methodological innovation harnesses expert's knowledge to enhance the predictive capabilities of maintenance strategies.

Previous studies have explored analogous methodologies in related domains. For instance, Boujelbene et al. utilized XGBoost, Bayesian ridge, and SVR techniques to optimize a novel Li-ion battery arrangement cooling system design [6]. Similarly, Duwon et al. undertook a comparative investigation contrasting physics-based modeling with neural network methodologies for forecasting cooling in integrated thermal management systems for vehicles [7]. Additionally, Wenxiu et al. developed a simplified mathematical model for vehicle engine electronic control cooling systems using the lumped parameter method. Their model integrated nonlinear and fuzzy control strategies to facilitate simulation [8].

To facilitate our analysis, we utilize advanced computational tools such as FisPro and OpenMarkov. FisPro, a specialized software for Fuzzy Inference Systems (FIS), facilitates the creation, simulation, and optimization of models, particularly suited for analyzing systems characterized by uncertainty and imprecision. Complementing this, OpenMarkov serves as a versatile platform for probabilistic graphical modeling, providing essential functionalities for model creation, manipulation, visualization, and probabilistic inference.

In this article, we present a comprehensive framework for our study, structured as follows: we begin with an introduction to the context and significance of our research, followed by a detailed exploration of Bayesian network construction. Subsequently, we delve into the calculation of probabilities within the model, followed by a rigorous validation process. We then apply the model to various scenarios and interpret the results. Finally, we conclude with a discussion of our findings and implications for future research endeavors.

2 Modeling the Failure Risk of Vehicles Cooling Systems Using a Fuzzy Bayesian Network Approach

In this study, we apply a fuzzy Bayesian network (FBN) methodology to assess the failure risk of vehicles cooling systems, as shown Fig. 1. The initial step involves identifying potential failure causes and determining the influencing parameters. We construct a BN, utilizing fuzzy logic (FL) to handle uncertain knowledge in the absence of precise events probabilities. By combining FL with experts' insights to establish rules and calculate conditional probabilities, we develop an FBN model that fuses these probabilities with the causal graph. This model enables the evaluation of failure probabilities under various conditions and simplifies the sensitivity analysis process.

2.1 Structure of the Bayesian Network

Bayesian networks are graphical models that represent variables as nodes and their conditional dependencies as directed edges in a structure known as a directed acyclic graph (DAG), ensuring there are no circular paths within the network [9]. The structure of the BN includes three principal elements [10]:

- Variables are depicted as nodes and are divided into three types: input, intermediate, and output.
- Directional links represent the dependencies among these variables.
- Conditional Probability Tables (CPTs) assign precise probabilities to these connections [11].

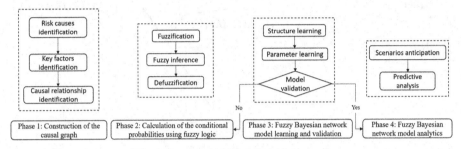

Fig. 1. Proposed methodology framework

As shown in Fig. 2, nodes $INPUT_1$, $INPUT_2$ and $INPUT_3$ are utilized as the parents of node $INTERMEDIATE_1$, exerting causal impact that affects it. In contrast, node $INTERMEDIATE_1$ acts as a child of the three initial inputs, its existence is the result of these three parent nodes. A BN may be structured to have either one output, in a configuration known as Multiple Input Single Output, or it can be designed with multiple outputs, in which case it is described as Multiple Input Multiple Output.

The analysis of diverse factors contributing to a cooling system's failure assists in determining the input parameters listed in Table 1, along with the intermediate effects and ultimate consequence. The causal graph depicted in Fig. 3

Figure 3 was formulated following a series of brainstorming sessions and experts' assessments to determine the crucial variables and their interdependencies.

Figure 3 illustrates the BN developed to model the factors that impact the vehicles cooling systems and cause its failure, created using the OpenMarkov software. In this representation, nodes are as follows: input nodes are described in Table 1, intermediate nodes are Radiator and cooling liquid, nodes with a direct impact on the risk of failure are intermediate nodes and thermostat and water pump, and the output node is the failure risk of cooling system. Additionally, the Radiator [12] is a unique node represents a dual role, functioning both as an input of cooling liquid and as an intermediate node that directly influences the risk of failure.

2.2 Calculation of Conditional Probabilities

Pursuing the steps for constructing the graph, it requires the calculation of CPs, which can be identified using two main approaches: objective methods and subjective methods. Objective methods rely on learning algorithms that utilize databases to calculate these

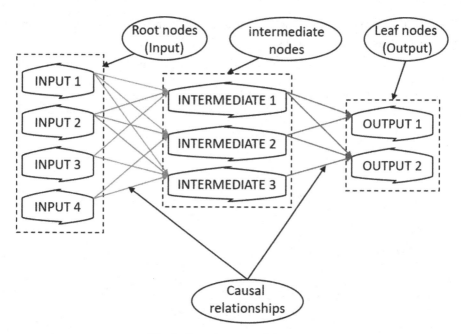

Fig. 2. Bayesian network structure

Table 1. Description of the input parameters of the causal graph.

Parameters name	Description
Thermostat [13]	it regulates the flow of coolant between the engine and radiator to maintain optimal engine operating temperature
Water Pump [14]	It pumps and circulates coolant throughout the engine and radiator to dissipate heat, helping to maintain proper operating temperature and to prevent overheating
Fan relay [15]	It controls the operation of the electric cooling fan(s), typically located near the radiator by activating and disactivating the electrical motor that turns the fans
Fan electrical fuse [15]	It prevents damage to the fan motor or other electrical components in the cooling system by interrupting the flow of electrical current in case of fault
Fan Motor [16]	It is responsible for driving the rotation of the electric cooling fan, which helps to increase airflow through the radiator
Head gasket [17]	It seals the combustion chambers, coolant passages, and oil passages in the engine to prevent all kinds of leaks

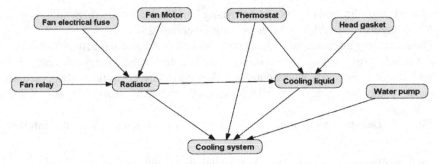

Fig. 3. Structure of Bayesian network modeling failure risk of vehicles cooling systems.

probabilities [11, 18, 19]. On the other hand, subjective methods involve the use of experts' knowledge in the field, often gathered through brainstorming sessions [10, 20–22]. Some studies employ a hybrid approach, leveraging both subjective and objective methods to enhance the accuracy and reliability of CP calculations [23]. As per this study, the subjective method was adopted by integrating the fuzzy logic system approach.

The fuzzy inference system is a process that calculates output parameters using a set of rules written in natural language. The implementation of this system requires three steps, as shown in Fig. 4, that constitute the fuzzy inference engine [24]:

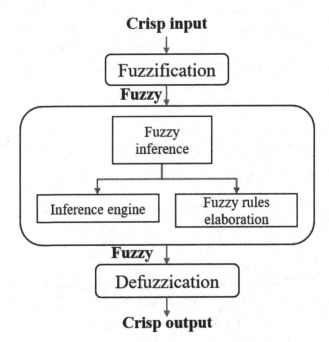

Fig. 4. A schematic explanation of Fuzzy Logic System

- **Phase 1:** Fuzzification involves transforming input variables into fuzzy subsets using fuzzy linguistic variables and membership functions [25].
- **Phase 2:** Fuzzy inference integrate fuzzy rules and membership functions to calculate the system's fuzzy output [25]. It is important to note that these fuzzy rules can follow the Mamdani approach, which considers the output variable as fuzzy (bad, medium, good), or the Sugeno approach, which regards the output as crisp (bad = 0.2, medium = 0.5, good = 0.8).
- **Phase 3:** Defuzzification transforms the fuzzy values generated by the inference engine into crisp values [26].

In the following part, these steps will be detailed in the context of our study.

- **Phase 1: Fuzzification.** For this phase, it is necessary to define four elements for each node: the variable's name, its linguistic values, its membership functions, and the universe of discourse. For this study, all nodes are characterized by three linguistic values, as shown in Table 2, utilize Gaussian membership functions, as shown Fig. 5, and operate within a range from 0 to 1.

Table 2. Linguistic values of our Bayesian Network

Nodes	Linguistic values	Nodes	Linguistic values
Fan relay	Good, Medium, Bad	Water pump	Work, Medium, Defective
Fan electrical fuse	Good, Medium, Bad	Radiator	Good, Medium, Bad
Fan motor	Good, Medium, Bad	Colling liquid	Full, Medium, Low
Thermostat	Work, Medium, Defective	Colling system	Good, Medium, Bad
Head gasket	Good, Medium, Damaged		

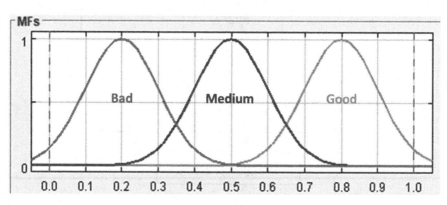

Fig. 5. Gaussian membership

- **Phase 2: Fuzzy inference.** This phase consists of 2 steps: Fuzzy logic elaboration and inference engine.

- The creation of fuzzy rules to determine final impacts relies on the expertise of specialists. These rules can be simple or intricate. Simple rules follow the format "IF X is A THEN Z is C," where X and Z are linguistic variables, and A and C represent their specific linguistic values. The format for complex rules is "IF X is A OP Y is B THEN Z is C," involving X, Y, and Z as the linguistic variables, and A, B, and C as their corresponding linguistic values. The operator OP, which can be "and" or "or," links the linguistic variables. An example of fuzzy rules applied to the "Cooling liquid" variable is presented in Table 3.
- In the fuzzy inference system, the inference engine implicates the antecedents to determine partial conclusions and subsequently aggregates the rules to establish the conclusion, employing Zadeh's operation (min for "or", max for "and"). Figure 6 shows the partial conclusions associated to the variable "Colling liquid", where the states of its antecedents are considered: Radiator "good", Thermostat "medium" and head gasket "good".

Table 3. Fuzzy rules of the "Cooling Liquid" node.

Rule	If Radiator	And Thermostat	And head gasket	Then Cooling liquid is
1	Bad	Defective	Damaged	0.2
2	Bad	Defective	Medium	0.2
3	Bad	Defective	Good	0.2
4	Bad	Medium	Damaged	0.2
5	Bad	Medium	Medium	0,5
6	Bad	Medium	Good	0,5
7	Bad	Work	Damaged	0,2
8	Bad	Work	Medium	0,5
9	Bad	Work	Good	0,8
10	Medium	Defective	Damaged	0,2
11	Medium	Defective	Medium	0,5
12	Medium	Defective	Good	0,5
13	Medium	Medium	Damaged	0,5
14	Medium	Medium	Medium	0,5
15	Medium	Medium	Good	0,5
16	Medium	Work	Damaged	0,5
17	Medium	Work	Medium	0,5
18	Medium	Work	Good	0,8
19	Good	Defective	Damaged	0,2

(*continued*)

Table 3. (*continued*)

Rule	If Radiator	And Thermostat	And head gasket	Then Cooling liquid is
20	Good	Defective	Medium	0,5
21	Good	Defective	Good	0,8
22	Good	Medium	Damaged	0.5
23	Good	Medium	Medium	0.5
24	Good	Medium	Good	0.8
25	Good	Work	Damaged	0.8
26	Good	Work	Medium	0.8
27	Good	Work	Good	0.8

The conclusion for this case, as shown in Fig. 6, is calculated as follows:

- Colling liquid (low) = 0.001
- Colling liquid (medium) = max (0.002, 0.135) = 0.135
- Colling liquid (high) = max (0.0.002, 0.607, 0.044) = 0.607

To utilize these values, it's essential that their sum equals 0, since they are conditional probabilities. Thus, a step of normalization is necessary to achieve this requirement. The calculation of these normalized values proceeds as follows:

- P (Colling liquid = low | Radiator = Good and Thermostat = Medium and Head gasket = Good) = 0.001/(0.001 + 0.135 + 0.607) = 0.001
- P (Colling liquid = medium | Radiator = Good and Thermostat = Medium and Head gasket = Good) = 0.135/(0.001 + 0.135 + 0.607) = 0.182
- P (Colling liquid = high | Radiator = Good and Thermostat = Medium and Head gasket = Good) = 0.607/(0.001 + 0.135 + 0.607) = 0.817

Following the same method, the remaining CPs for the "Cooling liquid" node are calculated and resumed in Fig. 7.

3 Model Validation, Scenarios Anticipation and Interpretation

3.1 Validation of Our Model

To validate the model proposed in this study, three axioms proposed by [27] were used:

- Axiom 1: Any increase or decrease in the probability of a parent node should result in a corresponding adjustment to the probability of its child node.
- Axiom 2: A variation in the probability distributions of a parent node should result in an impact on its child node.
- Axiom 3: The effect of each parent taken separately must be exceeded by the total effect of all parent nodes combined.

This finding for the axioms 1 and 2 for the node "Cooling liquid", and for axiom 3 for the node "Radiator", are presented in Tables 4 and 5 respectively.

Fuzzy Bayesian Network Applied to Modeling Vehicles Cooling Systems Failure Risk

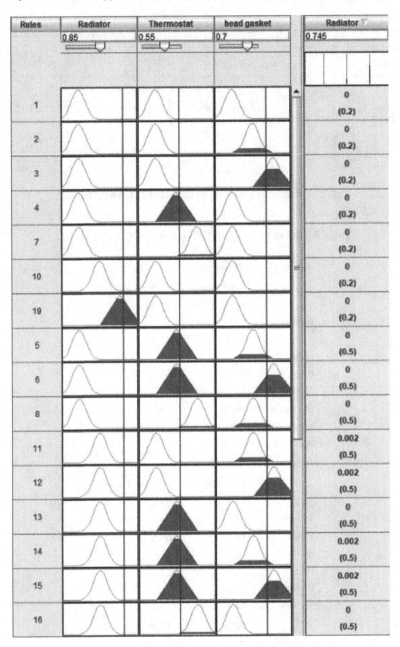

Fig. 6. Fuzzy inference of the node "Radiator".

Rule	Radiator	Thermostat	Head gasket	probabilities low	medium	high
1	0,2	0,15	0,1	0,995	0,003	0,002
2	0,2	0,15	0,5	0,987	0,012	0,001
3	0,2	0,15	0,7	0,981	0,018	0,002
4	0,2	0,55	0,1	0,981	0,018	0,002
5	0,2	0,55	0,5	0,012	0,976	0,012
6	0,2	0,55	0,7	0,002	0,931	0,067
7	0,2	0,75	0,1	0,981	0,018	0,002
8	0,2	0,75	0,5	0,012	0,976	0,012
9	0,2	0,75	0,7	0,001	0,182	0,817
10	0,5	0,15	0,1	0,995	0,003	0,002
11	0,5	0,15	0,5	0,012	0,976	0,012
12	0,5	0,15	0,7	0,017	0,965	0,017
13	0,5	0,55	0,1	0,017	0,965	0,017
14	0,5	0,55	0,5	0,012	0,976	0,012

Rule	Radiator	Thermostat	Head gasket	probabilities low	medium	high
15	0,5	0,55	0,7	0,017	0,917	0,066
16	0,5	0,75	0,1	0,017	0,965	0,017
17	0,5	0,75	0,5	0,012	0,976	0,012
18	0,5	0,75	0,7	0,001	0,182	0,817
19	0,85	0,15	0,1	0,995	0,003	0,002
20	0,85	0,15	0,5	0,012	0,976	0,012
21	0,85	0,15	0,7	0,001	0,182	0,817
22	0,85	0,55	0,1	0,003	0,003	0,993
23	0,85	0,55	0,5	0,002	0,950	0,047
24	0,85	0,55	0,7	0,001	0,182	0,817
25	0,85	0,75	0,1	0,002	0,067	0,931
26	0,85	0,75	0,5	0,001	0,047	0,951
27	0,85	0,75	0,7	0,002	0,067	0,931

Fig. 7. Conditional probabilities of the node "Cooling liquid".

Table 4. Test of axioms 1 and 2 for the node "Cooling liquid".

Axiom 1	Parent node	Child node
	Thermostat	Cooling liquid
20% increase	95%	77.94%
10% increase	85%	70.43%
Prior probability	75%	62.44%
5% decrease	70%	59.05%
10% decrease	65%	55.32%
Axiom 2	Parent node	Child node
	Head gasket	Cooling liquid
20% increase	90%	73.43%
10% increase	80%	65.67%
Prior probability	70%	57.56%
5% decrease	65%	53.52%
10% decrease	60%	49.56%

3.2 Scenarios Anticipation and Interpretation

After the Bayesian Network (BN) model is validated against established axioms, the final phase involves formulating scenarios to forecast the likelihood of failure in vehicles

Table 5. Test of axiom 3 for the node "Radiator".

Fan relay	Fan electrical fuse	Fan motor	Radiator	Percentage variations
70%	80%	90%	89%	0% (Initial value)
100%	80%	90%	92%	3.37%
70%	100%	90%	94%	5.61%
70%	80%	100%	96%	7.86%
100%	100%	100%	100%	12.36%

cooling systems. We manipulate the root nodes' input states to observe how changes affect the probability outcomes.

The scenarios are designed by categorizing input variables into Electrical Components (EC)—consisting of the Fan relay, Fan electrical fuse, and Fan Motor—and Liquid Circulation Components (LCC), which include the Thermostat, Water Pump, and Head Gasket. As outlined in Table 6, we consider scenarios for each category under both Favorable and Unfavorable conditions, culminating in four unique cases that assess the system's failure risk.

Table 6. Predictive scenarios for cooling system failure risk.

	Scenario 1	Scenario 2	Scenario 3	Scenario 4
EC	Unfavorable	Favorable	Unfavorable	Favorable
LCC	Unfavorable	Unfavorable	Favorable	Favorable

Our study, as in Fig. 8, underscores the pronounced influence of the Liquid Circulation Components (LCC) over the Electrical Components (EC) in determining the risk of failure within vehicles cooling systems. Through graphical analysis, we observe that unfavorable conditions within the LCC group correlate strongly with a heightened probability of system breakdown, reaching 80.3%. Conversely, unfavorable conditions within the EC group, juxtaposed with favorable LCC conditions, result in a substantially lower risk of failure, ranging between 0% and 15%. Consequently, our findings advocate for a strategic shift towards prioritizing monitoring and maintenance efforts on LCC components when forecasting cooling system failures. Such a targeted approach is poised to enhance the effectiveness of preventive measures and resource allocation, thereby contributing to the overall reliability and performance of vehicles cooling systems.

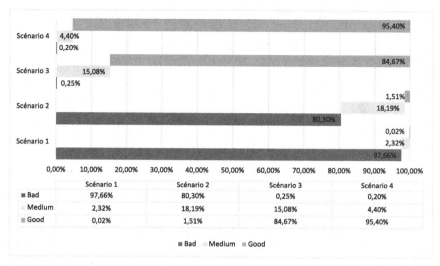

Fig. 8. Result of probability distribution for cooling system failure risk of each scenario

4 Conclusion, Discussions, Limitations, and Future Research

In conclusion, this study presents a hybrid approach utilizing a fuzzy Bayesian network to assess the failure risk of vehicles cooling systems. Through the construction of a Bayesian network model, we quantified the influence of root nodes on leaf nodes using a subjective method implemented with OpenMarkov software. Furthermore, fuzzy logic was employed to address uncertain decision-making, calculating conditional probabilities within the network using the Sugeno method implemented in FisPro software. To validate the model, we verified three axioms for all nodes in the constructed model. Subsequently, scenario analysis was conducted by categorizing input variables into two groups: Electrical components and Liquid circulation components. Each group was evaluated under both favorable and unfavorable conditions. The findings revealed that the highest probability of system failure was associated with unfavorable conditions within the Liquid Circulation Components (LCC) group, reaching as high as 80.3%. Conversely, the failure probability of Electrical Components (EC) under unfavorable conditions was limited, peaking at 15%.

These findings underscore the critical importance of prioritizing maintenance efforts towards mitigating risks associated with LCC components. Moving forward, our model provides a valuable framework for implementing proactive maintenance strategies aimed at optimizing system's performance and reducing downtime in industrial contexts.

This study acknowledges certain limitations that must be considered for a comprehensive understanding of its findings. Identifying all factors contributing to potential failures in the cooling system poses a challenge, and the subjective nature of establishing connections between variables, defining fuzzy sets, and determining membership functions may introduce inaccuracies due to the reliance on experts' judgments. To address these limitations and enhance future research, it is recommended to incorporate real-world data from industrial sources for more objective analysis.

Acknowledgment. The support for this research is provided by the Ministry of Higher Education, Scientific Research, and Innovation, as well as the Digital Development Agency (DDA) and the National Center for Scientific and Technical Research (CNRST) of Morocco, under the Smart DLSP Project - AL KHAWARIZMI AI-PROGRAM.

References

1. Meng, H., Li, Y.-F.: A review on prognostics and health management (PHM) methods of lithium-ion batteries. Renew. Sustain. Energy Rev. **116**, 109405 (2019)
2. Wen, Y., Rahman, Md.F., Xu, H., Tseng, T.-L.B.: Recent advances and trends of predictive maintenance from data-driven machine prognostics perspective. Measurement **187**, 110276 (2022). https://doi.org/10.1016/j.measurement.2021.110276
3. Ramesh, P.G., Dutta, S.J., Neog, S.S., Baishya, P., Bezbaruah, I.: Implementation of predictive maintenance systems in remotely located process plants under industry 4.0 scenario. In: Karanki, D., Vinod, G., Ajit, S. (eds.) Advances in RAMS Engineering. SSRE, pp. 293–326. Springer, Cham (2020). https://doi.org/10.1007/978-3-030-36518-9_12
4. Javaid, M., Haleem, A.: Impact of industry 4.0 to create advancements in orthopaedics. J. Clin. Orthop. Trauma **11**, S491–S499 (2020). https://doi.org/10.1016/j.jcot.2020.03.006
5. Haleem, A., Javaid, M.: Additive manufacturing applications in industry 4.0: a review. J. Ind. Integr. Manag. **04**(04), 1930001 (2019). https://doi.org/10.1142/S2424862219300011
6. Boujelbene, M., et al.: Machine-learning optimization of an innovative design of a Li-ion battery arrangement cooling system. J. Energy Storage **58**, 106331 (2023). https://doi.org/10.1016/j.est.2022.106331
7. Choi, D., An, Y., Lee, N., Park, J., Lee, J.: Comparative study of physics-based modeling and neural network approach to predict cooling in vehicle integrated thermal management system. Energies **13**(20), 5301 (2020). https://doi.org/10.3390/en13205301
8. Wu, W., Gui, C., Yang, P., Chen, W., Tan, Y.: Research on vehicle engine cooling system based on self-adjusting fuzzy control strategy. In: 2021 IEEE 4th International Electrical and Energy Conference (CIEEC), Wuhan, China, pp. 1–6. IEEE, May 2021. https://doi.org/10.1109/CIEEC50170.2021.9510562
9. Stephenson, T.A.: An Introduction to Bayesian Network Theory and Usage. Idiap (2000)
10. Bouhadi, O.E., Azmani, M., Azmani, A.: Using a fuzzy-Bayesian approach for predictive analysis of delivery delay risk. Int. J. Adv. Comput. Sci. Appl. **13**(7) (2022)
11. Chen, C., Liu, X., Chen, H.-H., Li, M., Zhao, L.: A rear-end collision risk evaluation and control scheme using a Bayesian network model. IEEE Trans. Intell. Transp. Syst. **20**(1), 264–284 (2019). https://doi.org/10.1109/TITS.2018.2813364
12. Paul Linga Prakash, R., Selvam, M., Alagu Sundara Pandian, A., Palani, S., Harish, K.A.: Design and modification of radiator in I.C. engine cooling system for maximizing efficiency and life. Indian J. Sci. Technol. **9**(2) (2016). https://doi.org/10.17485/ijst/2016/v9i2/85810
13. Ghasemi Zavaragh, H., Kaleli, A., Solmuş, İ., Afshari, F.: Experimental analysis and evaluation of thermostat effects on engine cooling system. J. Therm. Sci. **30**(2), 540–550 (2021). https://doi.org/10.1007/s11630-020-1264-8
14. Chastain, J., Wagner, J., Eberth, J.: Advanced engine cooling–components, testing and observations. IFAC Proc. **43**(7), 294–299 (2010)
15. Hoe, T.K.: Vehicle overheat prevention system: cooling fan failure alert system (2016)
16. Wang, T., Jagarwal, A., Wagner, J.R., Fadel, G.: Optimization of an automotive radiator fan array operation to reduce power consumption. IEEEASME Trans. Mechatron. **20**(5), 2359–2369 (2015). https://doi.org/10.1109/TMECH.2014.2377655

17. Park, M., Chang, T.: A study on automated tuning of the head gasket coolant passage hole of the gasoline engine cooling system using optimization technology. Presented at the Asia-Pacific Automotive Engineering Conference, no. 2019-01-1411, March 2019. https://doi.org/10.4271/2019-01-1411
18. Wu, B., Tang, Y., Yan, X., Guedes Soares, C.: Bayesian network modelling for safety management of electric vehicles transported in RoPax ships. Reliab. Eng. Syst. Saf. **209**, 107466 (2021). https://doi.org/10.1016/j.ress.2021.107466
19. Yan, L., Huang, Z., Zhang, Y., Zhang, L., Zhu, D., Ran, B.: Driving risk status prediction using Bayesian networks and logistic regression. IET Intell. Transp. Syst. **11**(7), 431–439 (2017). https://doi.org/10.1049/iet-its.2016.0207
20. Khalfaoui, H., Azmani, A., Farchane, A., Safi, S.: Symbiotic combination of a Bayesian network and fuzzy logic to quantify the QoS in a VANET: application in logistic 4.0. Computers **12**(2), 40 (2023). https://doi.org/10.3390/computers12020040
21. Khalfaoui, H., Azmani, A., Farchane, A., Safi, S.: Using a fuzzy-Bayesian approach for predicting the QoS in VANET. Appl. Comput. Syst. **27**(2), 101–109 (2022). https://doi.org/10.2478/acss-2022-0011
22. Benallou, I., Azmani, A., Azmani, M.: Evaluation of the accidents risk caused by truck drivers using a fuzzy Bayesian approach. Int. J. Adv. Comput. Sci. Appl. **14**(6) (2023)
23. Sharma, G., Bonato, M., Krishnamoorthy, M.: Bayesian network for reliability predictions of automotive battery cooling system. In: 2023 Annual Reliability and Maintainability Symposium (RAMS), Orlando, FL, USA, pp. 1–6. IEEE, January 2023. https://doi.org/10.1109/RAMS51473.2023.10088278
24. Bai, Y., Wang, D.: Fundamentals of fuzzy logic control—fuzzy sets, fuzzy rules and defuzzifications. Adv. Fuzzy Log. Technol. Ind. Appl., 17–36 (2006)
25. Martaj, N., Mokhtari, M.: MATLAB R2009, SIMULINK et STATEFLOW pour Ingénieurs, Chercheurs et Etudiants. Springer, Heidelberg (2010). https://doi.org/10.1007/978-3-642-11764-0
26. Théorêt, C.: Élaboration d'un logiciel d'enseignement et d'application de la logique floue dans un contexte d'automate programmable. École de technologie supérieure (2009)
27. Jones, B., Jenkinson, I., Yang, Z., Wang, J.: The use of Bayesian network modelling for maintenance planning in a manufacturing industry. Reliab. Eng. Syst. Saf. **95**(3), 267–277 (2010). https://doi.org/10.1016/j.ress.2009.10.007

Utilization of FMECA to Optimize Predictive Maintenance

Maria Eddarhri[1(✉)], Mustapha Hain[1], and Abdelaziz Marzak[2]

[1] ENSAM, University of Hassan II, Casablanca, Morocco
maria.eddarhri-etu@etu.univh2c.ma
[2] Faculty of Sciences, University of Hassan II, Casablanca, Morocco

Abstract. This paper explores the optimization of predictive maintenance in the aeronautical industry through the integration of Failure Mode, Effect, and Criticality Analysis (FMECA) with advanced technologies. It highlights the crucial importance of FMECA equipment in implementing predictive maintenance and shows how this approach can be used to identify and prioritize failures. By targeting high-risk failures, this innovative approach reinforces the predictive maintenance strategy by associating specific sensors. Integrating the Industrial Internet of Things (IIoT) and Artificial Intelligence (AI) is pivotal, enabling real-time equipment monitoring and data-driven decision-making. Moreover, the adoption of predictive maintenance from the manufacturing phase is emphasized as critical for manufacturers in a competitive environment, offering strategic advantages such as increased product reliability, reduced in-service failures, and lower maintenance costs. This not only enhances customer satisfaction and market reputation but also underscores a commitment to technological innovation and meeting market needs. Overall, this paper highlights an innovative approach to optimizing predictive maintenance in the aeronautical sector, aiming to reduce costs and improve safety through the strategic use of FMECA, IIoT, and AI. This synergy minimizes unplanned downtime, thereby enhancing operational efficiency and competitiveness in the aeronautical industry.

Keywords: Predictive maintenance · FMECA Equipment · Aeronautical industry · Industrial Internet of Things (IIoT) · Artificial Intelligence (AI)

1 Introduction

The aeronautical industry is renowned for its complexity and its constant quest for excellence in terms of safety, performance and reliability. At the heart of this quest is the maintenance of aeronautical equipment, a critical element in guaranteeing the smooth operation of aircraft. However, unexpected interruptions to production machinery have a direct impact on delivery times, ultimately affecting customer satisfaction and company profitability. This complex context has prompted a careful exploration of new approaches, and it is within this context that our paper takes its place. Our research focuses on the integration of Failure Mode, Effects and Criticality Analysis (FMECA) combined with Machine Learning for the implementation of predictive maintenance in the aeronautical industry.

FMECA is a proven methodology for assessing and managing the risks of machine failure and downtime. By integrating it with advanced Machine Learning techniques, we aim to revolutionize the way predictive maintenance is carried out in the aeronautical industry. This combination will enable more accurate prediction of failures, thereby reducing equipment downtime. In fact, Machine Learning will use the results of the FMECA Equipment to prioritize the failures to be dealt with in the process. In this paper, we focus on the development of a monitoring strategy for industrial equipment. The main objective is to identify and deploy optimal sensors to monitor the failure modes identified by the Failure Mode, Effect and Criticality Analysis (FMECA). This systematic method makes it possible to prioritize the risks associated with potential failures and to determine the most suitable sensors for detecting these anomalies before they lead to major failures. Through a detailed analysis, we examine the technical characteristics of the sensors, their accuracy and reliability, as well as their integration into the equipment's command and control system. The emphasis is on optimizing predictive maintenance to improve equipment availability, reduce maintenance costs and ensure safe operations. This paper therefore proposes a methodological approach for selecting sensors according to the specific needs identified by the FMECA, thus contributing to the continuous improvement of industrial maintenance processes. The objective is to provide a new perspective with regard to how FMECA combined with Machine Learning can revolutionize predictive maintenance in the aeronautical industry, enhancing operational safety while improving operational efficiency. Finally, sensors must be integrated in such a way as to facilitate their maintenance and calibration, thereby guaranteeing the reliability and accuracy of measurements over the long term. This systematic and strategic approach to sensor selection and installation is essential to optimizing the performance and safety of industrial equipment. It is important for manufacturers in a competitive world to consider predictive maintenance based on FMECA equipment at the production stage, in order to ensure effective monitoring and maintain a competitive lead in the market.

2 Review of Existing Approaches

Predictive maintenance has become crucial in industry to guarantee the safe and efficient operation of equipment. In this context, Failure Modes, Effects and Criticality Analysis (FMECA) is a well-established methodology for assessing the risks associated with industrial equipment failures. However, its integration with advanced technologies, such as artificial intelligence (AI) and the Internet of Industrial Things (IIoT), remains a constantly evolving area of research.

Paper [1] highlights the effectiveness of FMECA in diagnosing industrial equipment failures. It proposes an innovative approach using modified immune AI algorithms to improve the diagnostic process. This research demonstrates how FMECA can be enhanced with advanced techniques for more proactive equipment management. Another paper [2], goes even further by applying FMECA to Schneider Electric equipment. It shows how this methodology can be implemented in a real industrial context, demonstrating its feasibility and effectiveness in practical situations. This study provides a concrete example of the successful integration of FMECA into predictive maintenance. There is a study [3] that highlights the importance of safety risk assessment in IIoT environments, which is particularly relevant to the aeronautical industry. It highlights IEC 62443 standards as a framework for this assessment. This perspective is crucial, as it emphasizes that the integration of FMECA in aeronautics must take safety into account, aligning methodologies with sector-specific standards.

In summary, these papers provide a solid basis for understanding the importance of FMECA in predictive maintenance, its practical application in real industrial contexts and the importance of safety in IIoT environments. Indeed, our paper builds on this perspective by adapting this knowledge to the aeronautical industry, making a valuable contribution to existing research [4].

3 Materials and Methods

FMECA (Failure Modes, Effects, and Criticality Analysis) is an inductive framework that systematically dissects the origins and impacts of faults within system components, be it a product, machinery, or a procedural sequence. As a qualitative reliability technique, it enables the foresight of defect risks, alongside assessing their implications and pinpointing root causes [4]. Its principal focus is on cultivating systemic quality. FMECA endeavors to consolidate the collective skills of teams across the production spectrum to devise a suite of actions aimed at elevating the quality of products, processes, and the production environment at large.

Widely recognized as a fundamental tool in project oversight, maintenance, and comprehensive quality management, the FMECA protocol is adept at pinpointing failure triggers, delineating failure modes, and understanding their effects. The choice of a specific analytical approach is predicated upon the availability of pertinent data during the analysis [5, 6]. FMECA represents an advancement of the FMECA process, further integrating a criticality assessment that quantifies the probability and severity associated with any potential failure mode [4].

The FMECA approach is a specialized engineering analysis conducted by experts within a particular field, endowed with distinct capabilities such as:

- It enables the detection of equipment malfunctions and their origins and evaluates the fallout of these malfunctions on the system as a whole.
- It facilitates the evaluation of system failure risks and assists in prioritizing decision-making.
- It provides a means to address and rectify the most critical issues.
- It allows for the storage of analytical data in a central knowledge repository.

Figure 1 presents an overview of the tabular structure employed in the FMECA process.

The primary aim of this approach is to identify failures mode, their impacts and their criticity, subsequently arranging the failure modes in order of their associated risks [7]. FMECA evaluates each potential failure by assigning values based on Severity (S), Occurrence (O), and Detection (D), utilizing established rating scales. Table 1 depicts these scales for the classification of failure modes [8]. Broadly speaking, in any given risk assessment context, failure modes are sorted by their Risk Priority Number (RPN), which gauges the criticality of a failure mode as prescribed by the formula [9]:

$$RPN = S \times O \times D \tag{1}$$

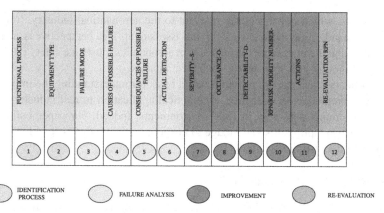

Fig. 1. FMECA methodology

Table 1. FMECA Rating Grid.

Severity (S)	Occurrence (O)	Detection (D)	Rating
Hazard without prior indication	Failure almost certain	Completely uncertain	10
Hazard with prior indication	Failure almost certain	Extremely unlikely	9
Extremely high	Frequent failures	Unlikely	8
Significantly high	Frequent failures	Very rare	7
Medium	Occasional failures	Rare	6
Slightly low	Occasional failures	Possible	5
Very minimal	Occasional failures	Fairly common	4
Minor hazard	Few failures	Common	3
Insignificant	Few failures	Very common	2
Negligible	Failure unlikely	Almost guaranteed	1

4 Proposed Approach

The FMECA equipment approach represents a logical evolution of our previous model that merges IIoT with artificial intelligence for optimized maintenance management [4]. In our previous research [4], we established a predictive model, based on sensor connectivity and real-time data collection, which uses machine learning algorithms to anticipate failures. FMECA reinforces this model by providing a structure for methodically assessing the severity, frequency and detectability of potential failures, thereby providing a comprehensive measure of RPN.

The FMECA equipment will enable us to specifically identify failure modes with high RPN. As part of this approach, we propose to associate them with sensors with a view to implementing predictive maintenance, thereby further strengthening our overall proactive maintenance management approach.

By performing a critical analysis of failure modes using FMECA and combining it with the accuracy of AI predictions based on IIoT data, we are significantly improving the accuracy of our predictive maintenance interventions. This synergy between FMECA and our AI-enhanced system enables us to accurately diagnose equipment operating conditions and prioritise maintenance actions according to the risks identified. Therefore, our FMECA equipment proposal goes beyond establishing a predictive maintenance model; it transforms our existing IIoT/IA model into an integrated and dynamic maintenance system, capable of adapting maintenance strategies in real time and optimizing production performance. Figure 2 illustrating the maintenance process using intelligent sensors, clearly demonstrates the interaction between FMECA equipment and our existing predictive model, highlighting the crucial role played by IIoT and AI in predicting failures and optimizing production lines.

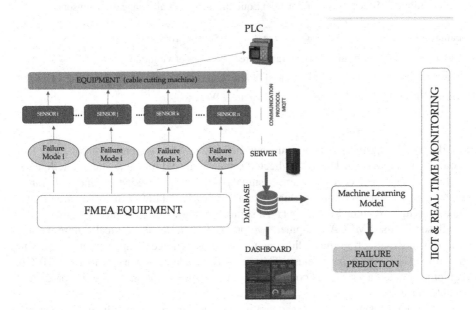

Fig. 2. Proposed Model

5 Results and Discussion

FMECA will enable us to specifically identify failure modes with a high RPN. As part of this approach, we propose to combine them with sensors with a view to implementing predictive maintenance, thereby further strengthening our overall proactive maintenance management approach.

Combining FMECA with IIoT encourages proactive management of equipment, minimizing unplanned downtime and extending its lifecycle. This synergy between the in-depth risk analysis of FMECA and the real-time monitoring capability of IIoT creates a more robust and responsive predictive maintenance system.

To illustrate, we use the example of a cable-cutting machine used in an industrial process. By applying the FMECA to this machine, we identify several potential failure modes, such as motor overheating, cutting blade breakage, power system malfunction and others. However, some of these failure modes can be particularly critical, such as excessive laser head vibration, bearing temperature and timing belt wear. To monitor these critical parameters, we carefully select the appropriate sensors. For instance, we install a vibration sensor on the laser head to detect any deviation from normal levels. Similarly, a temperature sensor is fitted to the bearing to signal any abnormal rise intemperature, which could indicate an imminent risk of failure. Finally, we use a wear sensor on the timing belt to monitor its condition and plan maintenance before it breaks, as shown in the Table 2.

Table 2. Specification of Potential Equipment Failures and Suggested Sensors.

Failure description	Failure type	Sensor proposed
Vibration of the laser head	Structural failure	Sensor of vibration
Temperature of the rolling	Mechanical failure	Sensor of temperature and sensor of noise
Wear of the distribution belt	Mechanical failure	Wear sensor

By integrating these sensors into our IIoT/IA system, we can monitor these key parameters in real time. The data collected is then analyzed by our artificial intelligence system, which can detect early warning signals of failure and predict impending failures. In this way, our cable-cutting machine becomes an essential link in the production chain, guaranteeing maximum efficiency and a significant reduction in unplanned downtime.

Based on the FMECA equipment we precisely identify the critical failure modes. The fundamental action that follows is the judicious choice of appropriate sensors to monitor these critical parameters. This process of adapting sensors to the FMECA equipment makes a significant contribution to the implementation of effective predictive maintenance.

It is crucial for manufacturers, operating in a demanding competitive environment, to adopt predictive maintenance based on FMECA equipment right from the manufacturing phase. This proactive approach offers a number of strategic advantages. Firstly, it ensures greater product reliability, reducing the risk of in-service failure and customer complaints. By minimizing unplanned interruptions to production, manufacturers can maintain a stable production flow, resulting in improved customer satisfaction and a positive reputation in the marketplace.

In addition, by implementing predictive maintenance at the manufacturing stage, warranty costs and maintenance expenditure can be reduced over the lifecycle of equipment. Manufacturers can therefore make substantial savings while offering high-quality products. This strengthens their competitiveness by enabling them to offer more reliable, high-performance equipment at competitive costs.

This approach also demonstrates the manufacturer's commitment to technological innovation and customer satisfaction. By integrating FMECA equipment and predictive maintenance right from the conceptual design phase, manufacturers are demonstrating their commitment to offering state-of-the-art products that are capable of meeting the changing needs of the market.

As a result, FMECA equipment -based predictive maintenance is much more than just a maintenance strategy; it is becoming an essential pillar of manufacturers' competitiveness, leading to better product quality, lower costs and a stronger reputation in the global marketplace.

6 Conclusion

In conclusion, this paper has demonstrated the compelling advantages of incorporating Failure Modes, Effects, and Criticality Analysis (FMECA) with advanced IIoT technologies to enhance predictive maintenance in the aerospace industry. By systematically identifying and prioritizing high-risk failure modes, and strategically deploying specific sensors, we can foresee and prevent equipment failures, thus significantly reducing unplanned downtime and extending the lifecycle of the equipment. The integration of FMECA equipment and IIoT technologies enables real-time monitoring and data analysis, fostering a more proactive and efficient maintenance management strategy. The use of predictive maintenance not only mitigates the risks of in-service failure and customer dissatisfaction but also solidifies the manufacturer's reputation for reliability and commitment to technological innovation.

Furthermore, adopting this approach from the manufacturing phase paves the way for substantial cost savings over the equipment's lifecycle and assures the delivery of high-quality products. This strategy underlines the manufacturer's dedication to excellence and positions them favorably in a competitive global market.

Ultimately, the synergy between FMECA equipment and IIoT is not merely a maintenance strategy but a transformative approach that propels the aerospace industry towards greater efficiency, safety, and competitiveness. As we continue to embrace these technologies, we set new standards for operational excellence and customer satisfaction in the aerospace sector.

To extend this research and explore its applicability beyond aerospace, future work will consider the automotive sector. The initial step will involve adapting FMECA and IIoT technologies to the unique characteristics of the automotive sector, where challenges and requirements may significantly differ. A comparative analysis of predictive maintenance between aerospace and automotive will provide valuable insights into the effectiveness and necessary adaptations for each sector. Moreover, integrating predictive maintenance into automotive quality management systems is crucial for enhancing reliability and customer satisfaction, underscoring the importance of this strategy for manufacturers. Finally, understanding the acceptability and adoption of predictive maintenance by consumers and manufacturers in the automotive industry will be essential to fully realize its transformative potential.

These future explorations aim not only to deepen our understanding of predictive maintenance but also to highlight its transformative potential across sectors, marking a significant advancement towards operational excellence and customer satisfaction across industries.

References

1. Samigulina et Samigulina: Diagnostics of industrial equipment and faults pre.pdf (2022)
2. Samigulina et Samigulina: Development of Industrial Equipment Diagnostics Sy.pdf (2020)
3. Hassani, et al.: Vulnerability and security risk assessment in a II.pdf (2021)
4. Barb, C.M., Fita, D.N.: A comparative analysis of risk assessment techniques from the risk management perspective. In: Bondrea, I., Cofaru, N.F., Inţă, M. (eds.) MATEC Web Conferences, vol. 290, p. 12003 (2019)
5. Barb, C.M.: A brief review of the risk assessment in Romania compared to Finland. In: Pasculescu, V.M. (ed.) MATEC Web Conferences, vol. 305, p. 00001 (2020)
6. Moraru, R.: A model for integrating work safety into the design of technical equipments. In: MATEC Web Conferences. Accessed 13 Jan 2024
7. Stamatis, D.H.: Failure mode and effect analysis - FMEA from theory to execution-American society for quality (ASQ) (2003). pdfcoffee.com. Accessed 13 Jan 2024
8. J1739_202101: Potential Failure Mode and Effects Analysis (FMEA) Including Design F. MEA, Supplemental FMEA-MSR, and Process FMEA - SAE International. Disponible sur. https://www.sae.org/standards/content/j1739_202101/. Accessed 13 Jan 2024
9. Anes, V., Henriques, E., Freitas, M., Reis, L.: A new risk prioritization model for failure mode and effects analysis. Qual. Reliab. Eng. Int. **34**, 516–528 (2018)
10. Eddarhri, M., Hain, M., Adib, J., Marzak, A.: A proposed roadmap for optimizing predictive maintenance of industrial equipment. Int. J. Adv. Comput. Sci. Appl. IJACSA (2023)

Comprehensive Study on Sentiment Analysis: Insights Into Data Preprocessing, Feature Extraction, and Deep Learning Model Selection

Ibtissam Youb[1,2(✉)] and Sebastián Ventura[2]

[1] CCPS Laboratory, ENSAM, University of Hassan II, Casablanca, Morocco
Yibtissam7756@gmail.com
[2] Department of Computer Science and Numerical Analysis, University of Córdoba, Córdoba, Spain

Abstract. Sentiment analysis, a pivotal task in natural language processing, is influenced by nuanced techniques in data preprocessing, feature extraction, and deep learning model selection. This research comprehensively explores the impact of varied methods, such as preprocessing techniques (steeming and stop-word removal), word embedding models (word2vec, Glove, and FastText), and deep learning architectures (DNN and RNN). Through rigorous evaluations on datasets including IMDB, SST2, and a collection of word tweets, the study identifies successful practices in sentiment classification. The ultimate goal is to offer practical recommendations for practitioners navigating sentiment analysis tasks with deep learning techniques.

Keywords: Sentiment analysis · Deep learning · Word embedding · Text preprocessing

1 Introduction

Opinions are at the heart of all human activities; they are the main factor influencing our behavior [7]. The opinions of others often determine how we think and act. In fact, When faced with a decision, we naturally seek the opinion of others; this is true not only for individuals, but also for businesses. The emergence of Web 2.0 has allowed people to exercise their right to express themselves freely and share their opinions through social media and other platforms. This has created more opinion data than ever before, which is why automated opinion research has become a necessity. Sentiment analysis is a fundamental technique for processing this data. Sentiment analysis can be defined as *"The automatic study of people's feelings, emotions, and attitudes toward entities such as products, services, and organizations"* [19]. Since the early part of this decade, sentiment analysis has caught the attention of many researchers, whether in computer science or in other fields such as management and social sciences [19]. Sentiment analysis is important for business and society as whole, and therefore It has evolved to be amongst the most important tasks in natural language processing and one of the most active areas in the field. Researchers are constantly developing, evaluating and comparing new techniques to solve the problem of opinion classification. Most of them adopt a two-step procedure: first, craft features are extracted from the text, and then these craft features are fed into a supervised machine learning algorithm. Popular choices for craft techniques include BOW and TF-IDF. Although these text vectorization models offer interesting results, they nevertheless suffer from serious limitations, as they do not consider the text's semantic or grammatical structure. To overcome these limitations, researchers have explored new approaches based on shallow neural networks. Neural approaches are based on an integration model that maps the text into a low-dimensional textual space. They use a deep learning algorithm to learn text representation from the context of the words. One of the earliest word embedding models is Word2Vec developed by [12] in 2013 and trained on 6 billion words. Subsequently, a range of word embedding models have been developed, including GloVe [15] developed by Stanford in 2014 and FastText [8] by Facebook in 2016. Sentiment classification is a challenging process that requires solving many NLP problems. Thus, many factors can directly impact the success of sentiment classification. In this research, we want to examine the extent to which popular approaches in natural language processing influence sentiment classification models. Our aim is to measure the impact of text preprocessing, word embedding and DL architecture on sentiment analysis systems. The comparative investigation will be performed on three different datasets, namely the well-known MIDB dataset, SST2 and a set of real-word tweets collected on the covid-19 vaccination. The choice of datasets with different characteristics will allow us to understand which commonly employed techniques work best for sentiment classification. The rest of this paper is structured as follows: Sect. 2 discusses related works briefly. Section 3 describes the core process of sentiment analysis, namely preprocessing techniques for text, word embedding, and some popular deep learning architectures. Section 4 presents the problematic, the datasets, and the evaluation metric that was utilized. Section 6 reports on the experimental setup, findings, and comments on the results.

2 Related Works

Sentiment analysis is a flourishing research area; it provi des necessary challenges and important applications for researchers. In this section, we will introduce some of the publications that focused on comparing different techniques for the task of sentiment classification. Early papers on sentiment analysis used all types of supervised machine learning algorithms to solve different NLP tasks. [2] Compared between SVM, Naïve Bias and Alternating Decision Tree (ADTree) and found that the SVM gives the most accurate results. [14] presented an early survey of the development of sentiment analysis. The study highlighted a number of limitations. Manly due to the use of handcrafted Features.

Sentiment Analysis with Deep Learning. Deep learning has become a powerful machine learning tool. And has produced breakthrough achievements in many areas such as natural language processing and computer vision. Deep learning architectures like RNN and its derivatives have outperformed conventional machine learning algorithms and have shown great ability to handle sequence data. Many scholars are interested in using deep learning for sentiment analysis. Especially with the development of word embedding techniques. Word embedding is in its core a machine learning algorithms that maps text into a low dimensional vector space. Differently from classical feature extraction techniques (BOW or TF-IDF) these models learn the numerical representation of words from the text. These systems transform text taking into consideration important information such as word order and grammatical structure of text. The comparison of different word embedding models for sentiment analysis has attracted many researchers [4] compared the efficiency of three fundamental DL architectures namely: DNN, RNN and CNN across different datasets using two different feature extraction techniques: word embedding and TF-IDF. The results show that the integration of word2vec model-based feature extractor contributes significantly to the efficiency and quality of inferences with htese models. Results also shows that the combination of word2vec and RNN gives the best results. However, in the paper the authors considered data preprocessing for granted, and did not verify its impact on the model, also they did not consider comparing between different word embedding systems. [5] presented a comparison of the performance of the models and the explainability of the models. They conclude that even the latest DL models provide excellent prediction quality. They suffer, however, from the problem of explainability because we do not understand how and in what way they outperform the others in the language classification task. [11] tested the effect of four popular word embedding models: Word2Vec, Crawl Glove, Twitter Glove, and FastText. Results show that Stanford's Twitter Glove provides the best accuracy. [17] compared between eight deep learning models across numerous datasets using two types of input levels. Results show that many factors can affect the accuracy of sentiment analysis including the volume of training data, the choice of deep learning architecture and the type of inputs. The study however did not consider the effect of the word vectorization models. [9] studied the impact of several NLP preprocessing techniques on sentiment analysis in big data, by using specific preprocessing techniques; the authors improved the accuracy of the Naïve Bayes classifier. However, the study did not consider recent deep learning models that provide better results. The impact of processing was also addressed by [1], the authors prove that the preprocessing steps impact directly the classification of machine learning algorithms specifically Naïve Bayes. However, in this paper, the volume of the dataset considered was very limited. In this paper, we carry out experiments to do an empirical study of deep learning. Our aim is to investigate the influence of different preprocessing techniques and different feature extraction techniques on the performance of two standard deep learning models, namely: DNN and RNN.

3 The Process of Sentiment Analysis

Typically, sentiment analysis is based on three fundamental steps [19]. The first is pre-processing. The objective of this step is cleaning the text and preparing it for the vectorization step. The vectorization step (word embedding) is intended to convert the documents into digital format (vectors); this step is essential for sentiment analysis. Finally, these vectors are injected into classification algorithms to finally classify the text into specific categories. Each of these steps is briefly detailed in this section.

3.1 Data Preprocessing

Text processing is a central element of a standard text classification approach. In the sentiment analysis literature, it is common to perform tokenisation, stop word removal and steeming. In text analysis, The process of breaking a piece of text into smaller pieces is known as tokenisation: Tokens are smaller units, such as words or phrases, tokenisation is a sort of text segmentation. Stop words are words that appear frequently in text documents and do not have a specific subject. Therefore, stop words are generally considered irrelevant in text classification literature and are eliminated before classification. Stop words are language-specific, as in the case of steeming. The purpose of the steeming step is to extract the radical form from the derived one. Since we assume that the radical and derived words are semantically identical, steeming is used extensively in text analysis. These steps are widely used in text classification and sentiment analysis in particular, without considering their impact on classification accuracy. In this paper, we investigate the influence of integrating two common preprocessing steps, namely steeming and stop word removal, on deep learning-based sentiment classification.

3.2 Word Embedding

Numerous deep learning approaches in natural language processing require word embedding as an input feature [19]. Word embedding is a text mapping technique in a low-dimensional embedding space. Unlike weighting techniques such as BOW and TF-Idf word embedding provides distrusted representations for words, which means that the information about a specific concept is distributed along the vector space, this helps the model encode important information about the context of each word. The core learning algorithm of these models is a deep learning algorithm that learns the vector representation from text.

Word2Vec. One of the first and most popular word embedding techniques is Word2vec [12]. Word2vec uses neural networks to learn distributed representations of words. The word2vec architecture is a class of two models:

Continuous Skip-Gram Model : Skip-gram utilizes a single word to make a prediction about a specific target context. The purpose of learning the Skip-gram is to generate continuous word representations that may be used to describe surrounding words in a particular text or phrase. Formally, given a series of words $w_1, w_2, w_3, ..., w_T$, Skip-gram's goal is to minimize log-likelihood function's error.

$$-\sum_{t=1}^{N} \sum_{-m \leq j \leq m, j \neq 0} \log P(w_{t+j} \mid w_t) \qquad (1)$$

With N being the number of words in the sequence and m being the size of the context.

Continuous Bag-of-Words Model: CBOW uses context to predict a target word by optimizing the log function in 2:

$$-\sum_{t=1}^{T} \log P(w_t \mid w_{t-m}, \ldots, w_{t-1}, w_{t+1}, \ldots, w_{t+m}) \qquad (2)$$

GloVe. GloVe [15] is unsupervised learning algorithms for generating word embedding. What glove does is represent a word by a high-dimensional vector trained on the context of the word. The pre-trained model of glove, which was trained on the Wikipedia corpus, is particularly used in academic research for its high performance. The objective function of GloVe is:

$$f(w_i - w_j, \widetilde{w}_k) = \frac{P_{ik}}{P_{jk}} \qquad (3)$$

with w_i being the vector of words corresponding to word i, and P_{ik} being the probability that the kth word appears in the context of the ith word.

FastText. The FastText approach developed by [8] is an extended version of Word2vec. The aim is to enhance vector representations of words for morphology-rich languages by a vectorial representation of n-grams nature. A word is then presented as the summation of the vectors of its n-grams. In concrete terms, instead of learning word vectors directly as input, FastText represents each word as a set of n-grams of characters. Similar to GloVe or Word2vec, FastText is a method for representing static words. Because these models are extremely computationally expensive. It is not easy to train them every time one starts a NLP task. Therefore, the research community has relied on existing pre-trained models. These models have been trained on huge datasets and are freely available online. In this paper, we compare the behavior of three fundamental pre-trained word embedding models on benchmark datasets, in order to understand how each word embedding technique performs with different preprocessing and deep learning architectures. Table 1 contains details of the pre-entrained models used in this project.

Table 1. Details of pre-pretrained word embedding

Word embedding	Dimension	Vocabulary
GloVe	50	2 million
FastText	300	1 million
Word2Vec	300	3 million

3.3 Deep Learning Classifiers

The mapping of words into a vector space is only the first step in sentiment analysis. The choice of classifier is equally important. Although machine learning-based models [6] have performed extremely well, many researchers agree that the integration of DL models has revolutionized the state of the art in a number of NLP applications. The combination of word integration and deep learning models has proven to be very effective for many NLP tasks ranging from synonym search to machine translation and sentiment analysis. Different deep learning architectures have been adopted for sentiment classification including DNN-based models, RNN-based models, etc. Since our goal is to verify the impact of preprocessing techniques and feature extraction methods on the sentiment analysis model, we will use two reference architectures in the field of text classification, namely the standard deep neural network model and the recurrent neural network.

Deep Neural Network. The network's initial layer is the input layer. It connects the input vector (discussed in Sect. 3.2) to the deep layers of the network. The network's depth is determined by the number of layers between input and output layers. The output layer has an activation function which is important for the classification of the input. The dimension of the last layer is equal to the number of classes. Figure 1 illustrates a completely connected deep neural network.

Recurrent Neural Network. [10] is a deep learning architecture. The special characteristic of this structure is that it uses a memory cell to process a sequence of entries. RNNs are able to memorise information over long sequences, making them highly suitable for NLP tasks. An RNN may be expressed with the following:

$$\mathbf{x_t} = \mathbf{F}(x_{t-1}, u_t, \theta) \qquad (4)$$

x_t represents current state of the cell at time t, and u_t represents the current input at step t. To be more explicit, if we integrate the parameters of the weights to formulate the 4 equation, we obtain:

$$\mathbf{x_t} = W_{rec}\sigma(x_{t-1}) + W_{in}u_t + b \qquad (5)$$

where $\mathbf{W}_{rec}$ is the weight of the recurrent matrix, $\mathbf{W}_{in}$ is the wight corresponding to the input, $\mathbf{b}$ is the bayes. Figure 2 shows an example of RNN's architecture,

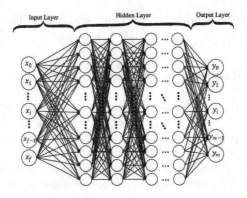

Fig. 1. Deep neural network (DNN) architecture [10]

the diagram on the left is the unzipped network, while the diagram on the right is the folded sequence neural network with tree time steps.

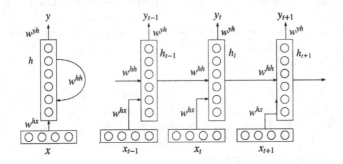

Fig. 2. Recurrent neural network [19]

Gated Recurrent Unit (GRU). RNN typically operates by either using the LSTM or the GRU for text classification. In this article, we use GRU which is an RNN blocking system developed by [3]. A detailed explanation of a GRU unit can be found below:

$$\mathbf{z_t} = \sigma_g \left(W_z x_t + u_z h_{t-1} + b_z \right), \tag{6}$$

W, U, and b represent component matrices and vectors, respectively. zt denotes the update gate vector at time t, xt. And the activation function is(σg).

$$\tilde{\mathbf{r}}_\mathbf{t} = \sigma_g \left(W_r x_t + u_r h_{t-1} + b_r \right), \tag{7}$$

with r_t is the reset gate vector at time t, z_t is update gate vector at time t.

$$\mathbf{h_t} = z_t \circ h_{t-1} + (1 - z_t) \circ \sigma_h \left(W_h x_t + u_h \left(r_t \circ h_{t-1} \right) + b_h \right) \tag{8}$$

where h_t indicates the output vector at time t, and (σ_h) is the hyperbolic tangent function.

4 Problem Formulation, Dataset, and Evaluation Metrics

Sentiment analysis is a challenging task. To extract sentiment from text, we need to go through the steps described in Sect. 3, namely: preprocessing, word embedding, and deep learning classification. Each of these steps contributes to the overall accuracy. The objective of this paper is to evaluate how changing the techniques used in each step impacts the performance and computational cost of the model. In the preprocessing step, we experimented with and without stop word removal and stemming, in the word embedding step, we tried three different embedding techniques, namely word2vec, glove, and FastText, and finally for the deep learning model, we examined RNN and DNN.

4.1 Datasets

The tests mentioned in this paper were carried out on three different databases:

IMDB: IMDB [13] is a reference database for sentiment classification. It is a set of movie reviews classified into two categories: Positive and Negative. Each review contains one or more sentences

SST2 : SSTb [18] The Stanford sentiment treebank is a database of movie reviews by Stanford University. www.rottentomatoes.com website. Unlike the IMDB dataset, SST2 is composed of short texts with only one sentence per review. The SST2 dataset used in this project is a binary dataset which consists of two classes: negative and positive.

Tweets Dataset: The Tweets dataset is a collection of tweets that we collected from Twitter on Covid-19 vaccination. We annotated them using the sentiment analysis tool Vader. Table 2 provides more details about the datasets.

Table 2. Dataset details

Dataset	Number of documents	Positive	Negative	Neutral	Average number of words per document	Max sequence long	Unique tokens	Unique tokens after preprocessing
IMDB	50000	25000	25000	-	225	2443	266708	210770
SST2	68221	38013	30208	-	9	48	26905	19664
TWEETS	30024	12044	11745	6235	29	59	32738	26077

Since these data were collected from the web, a data cleaning step was necessary. The purpose of this step is to remove noise from the text, for example in the case of Twitter data, removing user tags and HTML tags is paramount, similarly in the case of the IMDB dataset we removed the wristbands and HTML tags. Finally, we removed non-alphabetic characters in all three datasets.

4.2 Evaluation Metrics

To offer a better understanding of the performance of each model, two evaluation measures are considered:

Accuracy : is the number of correctly predicted examples out of the overall number of examples.

$$\text{Accuracy} = \frac{T_P + T_N}{T_P + T_N + F_P + F_N} \quad (9)$$

where: T_P represents the true positives, T_N are the true negatives, F_P are the false positives and F_N are the false negatives

F-score is a popular metric for classification models. It measures the model's accuracy on the dataset.

$$\text{F-score} = \frac{2T_P}{2T_P + F_N + F_P} \quad (10)$$

5 Experiments Results and Discussion

5.1 Experiments Setup

We have implemented the DNN and RNN architectures detailled in Table 3 using the Tensorflow and Keras platforms. The parameters of the models remained constant in all experiments: Epochs = 20, batch size = 1024. In addition, we uutilized 80% of each dataset as training data, and 20% as testing data. This approach will allow us to assess the impact of the various steps in the sentiment analysis. We used two evaluation measures, namely accuracy and F-score to evaluate our models. Finally, we ran the tests using the GPU provided by colab.

5.2 Experiments Results

Table 4 and 5 show the outcomes of the many experiments carried out in this study.

Table 3. Model layers details

	DNN	RNN
Layer-1	Embedding	Embedding
Layer-2	Flatten	GRU
Layer-3	Dense	dropout
Layer-4	dropout	GRU
Layer-5	Dense	dropout
Layer-6	dropout	Dense
Layer-7	Dense	

Table 4. Accuracy and F-score for different experiments without removing stop words and stemming

Dataset	embedding	RNN		DNN	
		Accuracy	F-score	Accuracy	F-score
IMDB	GloVe	87.32	87.32	85.65	85.65
	Word2vec	86.62	82.00	88.47	88.47
	FastText	88.64	88.64	87.87	87.87
SST2	GloVe	82.67	82.57	80.50	80.50
	Word2vec	82.11	82.04	79.01	78.93
	FastText	80.39	80.38	79.47	79.41
Tweets	GloVe	88.14	88.11	74.55	74.72
	Word2vec	81.52	82.63	74.15	74.20
	FastText	80.19	81.49	74.19	74.38

Table 5. Percentage of accuracy and F-score for the different experiments with removing stop words and stemming

Dataset	embedding	RNN		DNN	
		Accuracy	F-score	Accuracy	F-score
IMDB	GloVe	86.91	86.86	85.84	85.84
	Word2vec	84.64	84.62	88.41	88.41
	FastText	84.53	84.51	86.55	86.51
SST2	GloVe	77.64	77.60	75.80	75.67
	Word2vec	79.13	79.40	80.39	80.29
	FastText	76.72	76.49	77.06	76.95
Tweets	GloVe	78.83	78.84	71.31	71.16
	Word2vec	75.72	76.40	72.89	72.79
	FastText	77.46	78.30	71.53	72.00

5.3 Discussion

The combination of different preprocessing techniques, feature extractor and deep learning models produce different results across both metrics.

Deep Learning Models: The comparison between deep learning models, specifically DNN and RNN, sheds light on how they perform differently in sentiment analysis tasks.

RNN's Persistent Superiority: Consistently, RNN proves to be more effective than DNN, aligning with the broader sentiment analysis literature. RNN consistently achieves the highest accuracy and F-score across all datasets (88.64% - 82.67% - 88.14%), highlighting its ability to understand crucial sequential patterns essential for sentiment classification. This observation supports established studies that recommend RNN as the preferred model for sentiment analysis tasks [4].

Word Embedding: Exploring word embedding models, including Glove, FastText, and word2vec, alongside deep learning models offers insights into how they affect sentiment analysis outcomes.

Glove's Impressive Performance: Despite Glove having a smaller embedding dimension (50) compared to FastText and word2vec (300), it demonstrates noteworthy performance, particularly when paired with the RNN model. Statistical analyses confirm Glove's practical effectiveness, highlighting its ability to overcome the constraint of a smaller dimensionality. Achieving peak accuracy on datasets like SST2 (82.57%) and tweets (88.14%) positions Glove as a promising option for sentiment analysis tasks.

Varied Performance of FastText and word2vec: FastText and word2vec exhibit distinct strengths, showing variations in performance across datasets and models. Understanding these nuances requires a closer examination, especially concerning specific datasets and models. Future investigations may explore how embedding models interact with different deep learning architectures to identify optimal combinations.

Text Prepossessing: The linear regression analyses were conducted to scrutinize the influence of preprocessing techniques on model performance, specifically focusing on DNN accuracy and RNN accuracy. The obtained results illuminate nuanced relationships between these techniques and sentiment analysis outcomes [16].

DNN accuracy Model: The analysis of DNN accuracy reveals a modest impact of preprocessing techniques. The low R-squared value (0.01736) suggests that the chosen techniques explain only a small proportion of the variance in DNN accuracy. Furthermore, the non-significant p-value (0.602) for the Preprocessing technique coefficient indicates that the observed decrease in performance is not statistically significant. It is noteworthy that the negative coefficient suggests a potential adverse effect, although caution is warranted due to the lack of statistical significance.

RNN accuracy Model: Conversely, the analysis of RNN accuracy paints a more intricate picture. The higher R-squared value (0.2402) indicates that the chosen preprocessing techniques account for a more substantial portion of the variance in RNN accuracy. The statistically significant p-value (0.039) for the Preprocessing technique coefficient underscores a significant negative impact on the average performance of the RNN model. This suggests that, on average, the employed preprocessing techniques lead to a decrease in RNN accuracy. The negative coefficient aligns with the intuitive notion that certain preprocessing steps may hinder the model's ability to accurately classify sentiment.

These findings highlight the importance of carefully considering preprocessing techniques in the context of sentiment analysis with deep learning models. While the impact on DNN accuracy appears minimal and statistically non-significant, the significant negative impact on RNN accuracy warrants further investigation and consideration of alternative preprocessing strategies.

In assessing the impact of the different steps of the sentiment analysis process on the overall results, it is interesting to consider the computation time in addition to the quality of the prediction. For this purpose, we have reported the training time of each experiment in the Table 6.

Table 6. CPU time of the different experiments on the three datasets.

Dataset	embedding	RNN		DNN	
		ON	OFF	ON	OFF
IMDB	GloVe	1 h 16 min	1 h 13 min	32 s	31 s
	Word2vec	6 min 51 s	6 min 41 s	1 min 11 s	1 min 22 s
	FastText	25 min 24 s	21 min 30 s	2 min 5 s	2 min 30 s
SST2	GloVe	10 min 28 s	10 min 20 s	6 s	7 s
	Word2vec	6 min 03 s	11 min 36 s	17 s	21 s
	FastText	8 min 19 s	8 min 27 s	15 s	15 s
Tweets	GloVe	41 min 50 s	44 min 48 s	18 s	42 s
	Word2vec	3 min 41 s	4 min 26 s	21 s	12 s
	FastText	6 min 40 s	7 min 27 s	11 s	12 s

A notable point of observation is the substantial difference in training time between the DNN and RNN models. Despite its simplicity, the DNN model proves to be remarkably efficient, delivering optimal results in terms of both performance and computational cost. Importantly, the DNN model requires significantly less training time, making it a practical choice. The extended learning time for the IMDB dataset can be attributed to its large number of unique words and the high word count per document, as outlined in Table 2. The dataset's complexity contributes to the prolonged training duration, highlighting the computational challenges associated with handling sizable and intricate datasets. An interesting outcome from this study is that despite stemming and stop words removal reducing the overall number of words in the dataset, it does not notably improve computation time. This finding prompts a reconsideration of the necessity of these preprocessing steps for deep learning sentiment classification tasks. Therefore, omitting these steps could be a viable approach, especially when aiming to simplify the training process without compromising model performance.

6 Conclusion

In summery, our investigation scrutinized the dynamics of sentiment analysis, specifically exploring the effects of different stages on deep learning model performance. Notably, our findings urge a measured approach to common text preprocessing techniques like stop word removal and stemming, given their potential adverse impact on sentiment classification model quality. On a positive note, the endorsement of Glove for feature extraction is a standout recommendation. Despite its smaller vector dimension, Glove consistently showcased impressive performance across datasets. Additionally, our study reaffirms the anticipated superiority of recurrent deep learning architectures, emphasizing their potential to enhance the overall efficiency of sentiment analysis models. Looking forward, our suggestions advocate for a thoughtful balance in text preprocessing, strategic utilization of Glove, and a preference for recurrent deep learning architectures. As we anticipate further strides, the exploration of advanced deep learning architectures promises to play a pivotal role in the continual improvement and refinement of sentiment analysis models.

References

1. Alam, S., Yao, N.: The impact of preprocessing steps on the accuracy of machine learning algorithms in sentiment analysis. Comput. Math. Organ. Theory **25**(3), 319–335 (2019)
2. Annett, M., Kondrak, G.: A comparison of sentiment analysis techniques: Polarizing movie blogs. In: Conference of the Canadian Society for Computational Studies of Intelligence, pp. 25–35. Springer (2008)
3. Chung, J., Gulcehre, C., Cho, K., Bengio, Y.: Empirical evaluation of gated recurrent neural networks on sequence modeling. arXiv preprint arXiv:1412.3555 (2014)
4. Dang, N.C., Moreno-García, M.N., De la Prieta, F.: Sentiment analysis based on deep learning: a comparative study. Electronics **9**(3), 483 (2020)
5. Fiok, K., Karwowski, W., Gutierrez, E., Wilamowski, M.: Analysis of sentiment in tweets addressed to a single domain-specific twitter account: comparison of model performance and explainability of predictions. Expert Syst. Appl. **186**, 115771 (2021)
6. Hamlich Mohammed, M., Ramdani: Data classification by sac "scout ants for clustering" algorithm. J. Theoretical Appl. Inf. Technol. **55**, 66–73 (2013)
7. Ibtissam, Y., Abdallah, A., Mohamed, H.: Online panel data quality: a sentiment analysis based on a deep learning approach. Int. J. Artif. Intell. ISSN **2252**(8938), 1469 (2023)
8. Joulin, A., Grave, E., Bojanowski, P., Mikolov, T.: Bag of tricks for efficient text classification. arXiv preprint arXiv:1607.01759 (2016)

9. Khader, M., Awajan, A., Al-Naymat, G.: The impact of natural language preprocessing on big data sentiment analysis. Int. Arab J. Inf. Technol. **16**(3A), 506–513 (2019)
10. Kowsari, K., Jafari Meimandi, K., Heidarysafa, M., Mendu, S., Barnes, L., Brown, D.: Text classification algorithms: a survey. Information **10**(4), 150 (2019)
11. Krouska, A., Troussas, C., Virvou, M.: Deep learning for twitter sentiment analysis: the effect of pre-trained word embedding. In: Machine Learning Paradigms, pp. 111–124. Springer (2020)
12. Mikolov, T., Chen, K., Corrado, G., Dean, J.: Efficient estimation of word representations in vector space. arXiv preprint arXiv:1301.3781 (2013)
13. N, L.: Imdb dataset of 50k movie reviews, March 2019. https://www.kaggle.com/datasets/lakshmi25npathi/imdb-dataset-of-50k-movie-reviews
14. Nanli, Z., Ping, Z., Weiguo, L., Meng, C.: Sentiment analysis: a literature review. In: 2012 International Symposium on Management of Technology (ISMOT), pp. 572–576. IEEE (2012)
15. Pennington, J., Socher, R., Manning, C.D.: Glove: global vectors for word representation. In: Proceedings of the 2014 Conference on Empirical Methods in Natural Language Processing (EMNLP), pp. 1532–1543 (2014)
16. Saif, H., Fernández, M., He, Y., Alani, H.: On stopwords, filtering and data sparsity for sentiment analysis of twitter (2014)
17. Seo, S., Kim, C., Kim, H., Mo, K., Kang, P.: Comparative study of deep learning-based sentiment classification. IEEE Access **8**, 6861–6875 (2020)
18. Socher, R., Perelygin, A., Wu, J., Chuang, J., Manning, C.D., Ng, A.Y., Potts, C.: Recursive deep models for semantic compositionality over a sentiment treebank. In: Proceedings of the 2013 Conference on Empirical Methods in Natural Language Processing, pp. 1631–1642 (2013)
19. Zhang, L., Wang, S., Liu, B.: Deep learning for sentiment analysis: A survey. Wiley Interdisciplinary Reviews: Data Mining and Knowledge Discovery **8**(4), e1253 (2018)

A Combined AHP-TOPSIS Model for the Selection of Employees for Promotion

Loubna Bouhsaien[✉], Abdellah Azmani, and Imane Benallou

FST of Tangier, Abdelmalek Essaidi University, Tetouan, Morocco
loubna.bouhsaien@etu.uae.ac.ma, {a.azmani,ibenallou}@uae.ac.ma

Abstract. Human resources managers grapple with the intricate task of selecting employees for promotion, a process that often consumes several months and may be prone to unfairness. The fact that employee promotion is very low in the field of Decision Sciences was a call for this paper in order to help human resources managers, based on the combination of two Multi-Criteria Decision Methods (MCDM), namely the Analytical Hierarchy Process (AHP) and the Technique for Order Performance by Similarity to an Ideal Solution (TOPSIS), to systematically choose and rank the most appropriate candidates for promotion. Initially, AHP determined the importance of diverse criteria through pairwise comparisons by elaborating the weights. Subsequently, employees were randomly chosen from a pool of potential candidates seeking promotion, and TOPSIS is used to rank the employees based on these weighted criteria. The findings indicates that reputation, Key Performance Indicator (KPI), convergence with current work, and the performance ratings of the last year emerged as the top four variables influencing the promotion decision-making process.

Keywords: Employee's promotion · MCDM · AHP · TOPSIS

1 Introduction

Human Resources Management (HRM) has a fundamental role in the development of a company, emphasizing the importance of effective implementation and qualified work abilities among employees [1]. It advocates for a well-organized HRM system to facilitate the achievement of organizational goals [2].

Organizations must recognize that employees are the vital force driving their success [3]. In the age of globalization, advanced technology, and modern communication systems, it is crucial for any business aiming to stay competitive and operational to consistently seek strategies for attracting and retaining high-quality personnel [4–6].

For both organizations and individuals, promotion plays a significant role [7], fostering commitment and motivation among employees while enhancing their morale, well-being, and life satisfaction. However, alongside the benefits, promotion entails increased responsibilities, longer working hours, heightened stress levels, and a potential reduction in work-life balance [8]. A promoted employee is also expected to acquire familiarity and master new tasks.

The pie chart (Fig. 1) shows that the largest percentage of papers published in Scopus according to employee promotion is in the Medicine field, at 27.3%. The next largest percentage is in the Business, Management, and accounting field, at 14.3%. The Social Sciences and Humanities field is third, at 14.7%. It also shows that the percentage of papers published in Scopus according to employee promotion is very low in some fields, such as Decision Sciences and Information Systems (2.2%) and Nursing (3.9%).

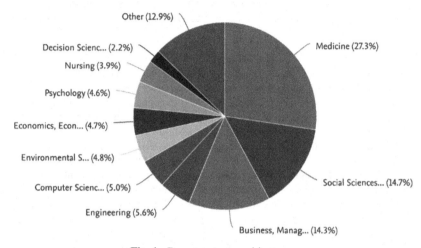

Fig. 1. Documents per subject area

The paper will start by a literature review, followed by an overview of the AHP and TOPSIS. The subsequent sections will focus on the practical application of the AHP-TOPSIS method and a discussion of the obtained results. This study aims to present a concise and effective exploration of the research process and its outcomes according to employees' promotion to help human resources managers choosing the most adequate employee.

2 Related Works

Building upon the foundations laid by prior research, the pursuit of equitable and efficient employee promotions emerges as a formidable challenge. In the present era, there is a consensus amongst researchers and managers that focusing on the perspectives of employees contributes to establishing a sustainable competitive advantage within organizations [9]. Previous studies have paved the way, and as we navigate the landscape of the Profile Matching Method and delve into the fascinating realm of Machine Learning techniques, we unravel the intricate interplay between data and decision-making in the corporate sphere. Additionally, the examination of job promotions through mobile sensing provides a unique lens into the physiological and behavioral dynamics influenced by these professional advancements.

The greatest challenge for human resources managers is to promote in a fair manner, knowing that humans generally rely on feelings. To address this issue, several studies have been conducted based on various factors. Table 1 shows an overview of previous studies on employee's promotion.

Table 1. Overview on some previous studies about employee's promotion

Methodology	Main idea
Decision support system – Profile Matching Method [10]	This method facilitates the creation of a decision support system, streamlining the selection process for specific positions based on the necessary criteria. The variables used in this methodology are broadly classified into two categories: those related to work attitude and others associated with the intellectual aspect
LR, RF, and AdaBoost [11]	A supervised learning approach was adopted to predict employee promotions using labeled historical data. They performed feature engineering, data splitting with five-fold cross-validation, addressed class imbalance with SMOTE, conducted hyperparameter tuning using grid search, and identified influential features through experiments to enhance prediction accuracy
SVM, RF, KNN, DT, LR, and AdaBoost [12]	To promote employees, two criteria were used: individual fundamental data, encompassing personal details, alongside position-related information employing six classification algorithms
LR, DT, RF, KNN, SVM, Gaussian Naïve Bayes, AdaBoost, XGB with SMOTE+ENN [13]	The study employs machine learning and a hybrid sampling approach to predict employee promotions, training and evaluating eight classifiers. Feature analysis highlights key attributes, with SMOTE+ENN and XGB using eight features identified as the best-performing combination
J48, random forest, naïve Bayes, and SMO [14]	Several factors were considered in selecting the most suitable employees for promotion, including education, training and development, language skills, performance evaluation, professional experience, interview and interpersonal skills, research and publication, and computer skills

(*continued*)

Table 1. (*continued*)

Methodology	Main idea
Fuzzy Logic [15]	The objective was to address the employee promotion process by evaluating factors such as ability and loyalty. The scoring rules, ranging from excellent (5) to the worst (1), were analyzed using the Mamdani Fuzzy method. The study concluded with a testing data accuracy of 91.4% and an average processing speed of 1.24 s
Mobile Sensing [8]	Examining data from phones, wearables, and Bluetooth beacons before and after promotions reveals changes in physiological and behavioral patterns. Metrics cover various aspects, including workplace activity, interactions, distance traveled, stress levels, sleep patterns, physical activity, heart rate, and phone usage

The methodologies employed are complex due to the complexity involved with data, nodes/variables, dependencies, and values, leading to high costs. AHP-TOPSIS presents a methodical framework for organizing criteria and efficiently assessing alternatives, based on experts, enabling the simultaneous consideration of multiple dimensions with ease and flexibility. In a professional approach, it is recommended to combine both starting with AHP and then utilizing the other methods to enhance decision-making.

3 Presentation of the AHP-TOPSIS Methods for Employees' Promotion

The AHP-TOPSIS method stands out as a robust and eligible tool for making effective decision-making scenarios characterized by multiple competing factors. The method initiates the decision-making process by facilitating pairwise comparisons of criteria, enabling decision-makers to assign scores that capture the relative importance of each criterion. Subsequently, the methodology engages in a meticulous calculation of the criteria's relative importance, establishing a hierarchy that reflects their significance in the decision context.

Moving forward, the AHP-TOPSIS (Fig. 2) method undertakes the normalization and weighting of scores attributed to each decision option based on the established criteria. This normalization process ensures that diverse factors, potentially with varying units or scales [16–18], are harmonized for a fair assessment. The subsequent weighting procedure incorporates the relative importance of each criterion, allowing for a comprehensive evaluation.

The distinguishing feature of the AHP-TOPSIS method lies in its identification of two crucial scenarios: the best and worst possible outcomes. By determining these ideal

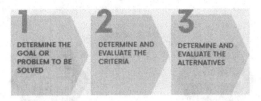

Fig. 2. Overview of AHP-TOPSIS method

scenarios, the methodology sets a benchmark against which the performance of each decision option is measured. This step is integral to the subsequent calculation of each option's closeness to the ideal best outcome.

The measure of closeness provides a nuanced and balanced perspective, revealing how well each decision option aligns with the ideal best scenario, considering all the competing factors at play. This structured and systematic approach empowers decision-makers with a holistic understanding of complex choices, fostering confidence in the decision-making process. By combining the strengths of pairwise comparisons, normalization, weighting, and the identification of ideal scenarios, the AHP-TOPSIS method becomes an invaluable ally in navigating complicated decision landscapes with clarity and precision.

4 Implementation of the AHP-TOPSIS Approach for Employees' Promotion

This section focuses on the application of the AHP-TOPSIS approach to evaluate criteria related to employee promotion.

Figure 3 provides an overview of the applied methodology adopted to the promotion challenge discussed in this paper. The process starts with the application of the AHP, involving the identification of predictors, pairwise comparisons through a matrix, and criteria weighting. Subsequently, the employee ranks are determined using the TOPSIS. This entails establishing scores for the selected criteria for each employee, constructing a weighted normalized matrix, and identifying both the best ideal and worst ideal solutions.

4.1 AHP

AHP is used in order to determine the importance of each criteria by calculating its weight [19].

- Step 1: Determine predictors.

Selecting candidates for promotion involves carefully assessing various factors. This study specifically looks at three key aspects: personal factors, work experience, and employee scores. These factors are central to determining the suitability of individuals for advancement within the organization.

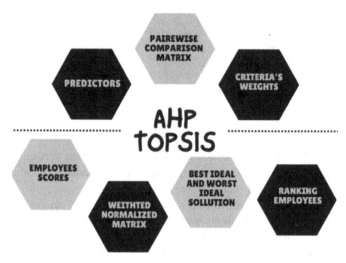

Fig. 3. AHP-TOPSIS Method Applied for Employee's Promotion

Identifying promotion sub criterion of each factor is a fundamental step in the employee promotion selection process. Based on the literature, databases, and brainstorming with experts, nine factors relating to employees impact the promotion were chosen as shown in Table 2. The variables reputation, convergence, training score, previous year rate, and KPI are score out of 10; the other variables are done by a number indicating the real value of the variable.

Table 2. Employee Promotion Criteria

Criteria	Sub criterion	Comment	Abbreviation
Personal factors	Reputation [21]	The perceived standing, regard, or esteem of an individual within the organization based on their professional conduct, achievements, and interactions	REP
	Age	The employee's age	AGE
Work experience	Years spent at the company [22]	The duration, measured in years, that an individual has been employed by a specific company or organization	Y_C

(*continued*)

Table 2. (*continued*)

Criteria	Sub criterion	Comment	Abbreviation
	Convergence of the current post	The degree of alignment or harmonization between the current post and the post exigence, indicating how closely the current situation corresponds to the required or desired state	CV
	Years with current manager [22]	The duration, measured in years, that an individual has been working under their current manager within a specific organization	Y_M
	Total working years [23]	The cumulative work experience period an individual has spent in employment or engaged in professional activities	Y_T
Employee scores	Average Training score [24]	The average score attained by an employee across various training sessions. It serves as a metric to assess the effectiveness of training programs tailored to organizational demands	TR
	Previous year rate [25]	The score given to an employee based on his performance during the previous year	RATE
	KPI [26]	A Key Performance Indicator (KPI) serves as a metric or vital indicator employed to assess the performance of an organization, department, or individual in attaining predetermined goals and objectives	KPI

- Step 2: Construct the pairwise comparison matrix.

Before delving into the construction of the pairwise comparison matrix, it is imperative to have a clear understanding of the numerical values that will be used in this matrix and their significations. The significance of the numerical values in the pairwise comparison matrix is detailed by Table 3. Each value represents a specific level of importance, ranging from equal importance 1 to extreme importance 9, with intermediate values; 2, 4, 6, 8; indicating gradations between adjacent levels [20].

To illustrate the application of this matrix, Table 4 summarizes the relationships between variables based on the input from domain experts. For example, KPI is considered four times more important than the convergence of the current role.

Table 3. Significance of the numerical value of the pairwise comparison matrix

Numerical value	Description
1	Equal importance
3	Slight importance of one over another
5	Moderate importance of one over another
7	Very strong importance
9	Extreme importance of one over another
2,4,6,8	Intermediate values between two adjacent values

Table 4. Pairwise matrix

	KPI	REP	TR	RATE	Y_C	CV	Y_M	Y_T	AGE
KPI	1	0,50	6,00	3,00	3,00	**4,00**	6,00	6,00	8,00
REP	2,00	1	4,00	3,00	4,00	5,00	4,00	6,00	4,00
TR	0,17	0,25	1	0,50	0,50	0,20	3,00	0,50	4,00
RATE	0,33	0,33	2,00	1	2,00	0,33	4,00	4,00	6,00
Y_C	0,33	0,25	2,00	0,50	1	0,50	3,00	5,00	6,00
CV	0,25	0,20	5,00	3,00	2,00	1	5,00	3,00	5,00
Y_M	0,17	0,25	0,33	0,25	0,33	0,20	1	0,50	3,00
Y_T	0,17	0,17	2,00	0,25	0,20	0,33	2,00	1	2,00
AGE	0,13	0,25	0,25	0,17	0,17	0,20	0,33	0,50	1

The values in the matrix represent the pairwise comparisons of each factor against every other factor. For example, the value at the intersection of KPI and REP signifies the importance of KPI relative to REP, and so on. This structured approach aids in capturing the nuances of expert opinions and lays the foundation for subsequent steps in our analysis.

- Step 3: Calculate the weight of each criterion.

To calculate the weight, each element was divided by the sum of its column just before calculating the average of all the rows (2). The obtained weights are presented by Table 5.

$$r_{ij} = \frac{x_{ij}}{\sqrt{\sum_{i=1}^{n} x_{ij}}} \quad (1)$$

$$w_{ij} = \frac{1}{n}\sqrt{\sum_{i=1}^{n} r_{ij}} \quad (2)$$

To verify the consistency of the comparison matrix, it is necessary to find that the consistency ratio (CR) is lower than 0,1. To do so, it is necessary to calculate the consistency index (CI) and find the appropriate random index (RI).

To calculate the ratio (3), it is necessary to sum all the elements of each column in the matrix and divide it by the associated weight to the criteria.

$$\text{ratio} = \sum_{j=1}^{n} \frac{x_{ij}}{weight_j} \quad (3)$$

Table 5. Criterion's normalization and weights

	KPI	REP	TR	RATE	Y_C	CV	Y_M	Y_T	AGE	Weight
KPI	0,22	0,16	0,27	0,26	0,23	0,34	0,21	0,23	0,21	0,23
REP	0,44	0,31	0,18	0,26	0,30	0,42	0,14	0,23	0,10	0,27
TR	0,04	0,08	0,04	0,04	0,04	0,02	0,11	0,02	0,10	0,05
RATE	0,07	0,10	0,09	0,09	0,15	0,03	0,14	0,15	0,15	0,11
Y_C	0,07	0,08	0,09	0,04	0,08	0,04	0,11	0,19	0,15	0,09
CV	0,06	0,06	0,22	0,26	0,15	0,08	0,18	0,11	0,13	0,14
Y_M	0,04	0,08	0,01	0,02	0,03	0,02	0,04	0,02	0,08	0,04
Y_T	0,04	0,05	0,09	0,02	0,02	0,03	0,07	0,04	0,05	0,04
AGE	0,03	0,08	0,01	0,01	0,01	0,02	0,01	0,02	0,03	0,02

Table 6 presents the Criterion's Ratio, offering a quantitative depiction of the relative importance assigned to each criterion in the evaluation process.

Table 6. Criterion's ratio

Criterion	KPI	REP	TR	RATE	Y_C	CV	Y_M	Y_T	AGE
Ratio	10,8	10,78	9,58	9,99	10,02	10,57	9,44	10,04	9,51

The value of λ_{max} is done by the average of all the ratios (4) calculated in Table 6.

$$\lambda_{max} = \frac{1}{n}\sum_{i=1}^{n} \text{ratio} = \frac{90{,}72}{9} = 10{,}08 \qquad (4)$$

The consistency index is then calculated as shown in (5).

$$CI = \frac{\lambda_{max} - n}{n - 1} = \frac{10{,}08 - 9}{9 - 1} = 0{,}135 \qquad (5)$$

Table 7 establishes a Random consistency index RI considering the number of criteria addressed by the hierarchy in question [18]. The RI corresponding to the number of criteria in our matrix is determined to be 1,45.

Table 7. RI Index Table

n	1	2	3	4	5	6	7	8	9	10
RI	0	0	0,58	0,9	1,12	1,24	1,32	1,41	1,45	1,49

Now, to calculate the consistency ratio, it is necessary to divide CI by RI as shown in (6).

$$CR = \frac{CI}{RI} = \frac{0,135}{1,45} = 0,092 \qquad (6)$$

Regarding the calculation, it is evident that CR < 0.1, affirming the consistency of the comparisons in the matrix.

4.2 TOPSIS

TOPSIS was used to rank the different alternatives, in other words, the employees, depending on the predetermined criterion [18, 27].

- Step 4: Determine the scores of each row of the matrix.

To apply TOPSIS, the first thing is to determine the scores of each criterion for all employees. Table 8 shows the scores (x_{ij}) of four employees that are the most eligible for the promotion.

- Step 5: Normalize the matrix and weight matrix.

To normalize the matrix, it is necessary to multiply the matrix elements by the weights already found (7). Table 9 shows the normalization results.

$$R_{ij} = w_{ij} \times x_{ij} \qquad (7)$$

After applying the Eq. (8), the weighted matrix is obtained and shown in Table 10.

$$r_{ij} = \frac{x_{ij}}{\sqrt{\sum_{i=1}^{n} x_{ij}^2}} \qquad (8)$$

Table 8. Employee's scores

Criteria	Employee 1	Employee 2	Employee 3	Employee 4
KPI	8	6	5	6
REP	5	8	6	5
TR	6	7	8	5
RATE	7	5	6	7
Y_C	5	4	6	4
CV	1	1	0	1
Y_M	6	1	5	6
Y_T	6	2	8	6
AGE	43	27	37	40

Table 9. Normalized matrix

Criteria	Employee 1	Employee 2	Employee 3	Employee 4
KPI	0,57	0,34	0,4	0,45
REP	0,43	0,55	0,47	0,32
TR	0,36	0,41	0,54	0,39
RATE	0,43	0,34	0,33	0,45
Y_C	0,57	0,34	0,4	0,45
CV	0,43	0,55	0,47	0,32
Y_M	0,36	0,41	0,54	0,39
Y_T	0,43	0,34	0,33	0,45
AGE	0,57	0,34	0,4	0,45

- Step 6: Variation from the positive and negative solution

In this step, the objective is to evaluate the variation of each criterion from both the positive (V+) and negative (V−) solutions.

Table 11 displays the calculated variations for each criterion. The values in the V+ column represent the maximum positive variation for each criterion (9), while the values in the V− column represent the minimum negative variation (10).

$$V_i^+ = \max(r_{ij}) \quad (9)$$

$$V_i^- = \min(r_{ij}) \quad (10)$$

The following step is about calculating the separation of alternatives from the positive and negative ideal solution.

Table 10. Weighted matrix

Criteria	Employee 1	Employee 2	Employee 3	Employee 4	Weight
KPI	0,13	0,09	0,02	0,05	0,23
REP	0,10	0,14	0,03	0,04	0,27
TR	0,08	0,11	0,03	0,04	0,05
RATE	0,10	0,09	0,02	0,05	0,11
Y_C	0,13	0,09	0,02	0,05	0,09
CV	0,10	0,14	0,03	0,04	0,14
Y_M	0,08	0,11	0,03	0,04	0,04
Y_T	0,10	0,09	0,02	0,05	0,04
AGE	0,13	0,09	0,02	0,05	0,02

Table 11. Variation from the positive and negative solution

Criterion	KPI	REP	TR	RATE	Y_C	CV	Y_M	Y_T	AGE
V+	0,13	0,14	0,03	0,05	0,05	0,07	0,02	0,01	0,01
V−	0,08	0,09	0,02	0,04	0,04	0,00	0,00	0,03	0,01

It is done based on the Euclidean distance using the Eqs. (11) and (12).

$$S_i^+ = [\sum\nolimits_{i=1}^{n} (R_{ij} - V^+)^2]^{1/2} \quad (11)$$

$$S_i^- = [\sum\nolimits_{i=1}^{n} (R_{ij} - V^-)^2]^{1/2} \quad (12)$$

Table 12. Separation of employees from ideal solution and rank

Employee	S+	S−	P
Employee 1	0,0033	0,0079	O,70
Employee 2	0,0020	0,0099	0,46
Employee 3	0,0092	0,0079	0,64
Employee 4	0,0047	0,0083	0,83

5 Discussion

To ensure a robust and systematic evaluation, a two-step methodology was employed. Firstly, the AHP method was used so that the weights of criteria were calculated. Secondly, the TOPSIS method was applied so that the ideal solutions were determined, for each alternative, the deviations were evaluated, and the coefficients were assigned for each employee to determine their ranks, culminating in a coherent decision-making process.

The employees are ranked based on their proximity to the ideal solution. As a result, employee 2 emerges as the prime candidate for promotion, attaining the highest rank of 0.83, while Employee 1 and Employee 4 follow with ranks of decreasing order, concluding with the lowest rank of 0.46 for Employee 3 as shown in Table 12.

This approach ensures a balanced evaluation [28], aligning with scientific principles and fostering a logical framework that contributes to an equitable promotion process.

6 Conclusion

This paper presents the utilization of a hybrid method, combining AHP and TOPSIS, to facilitate, for human resources managers, the decision-making process for selecting one amongst four employees to get a promotion.

The findings of the study reveal that several factors significantly influence the promotion decision-making process. Reputation, Key Performance Indicators (KPIs), alignment with current responsibilities, and past performance ratings emerge as the top four variables in this regard.

While these methods prove to be user-friendly and aid in decision-making with multiple alternatives, limitations arise in the determination of the values of the different parameters, especially for the comparison matrix in the AHP method [18].

To fortify this approach, the prospect of integrating artificial intelligence [29] and adopting an approach like Benefit, Opportunity, Cost, and Risk is considered [17, 30, 31].

Acknowledgments. This research is supported by the Ministry of Higher Education, Scientific Research and Innovation, the Digital Development Agency (DDA) and the National Center for Scientific and Technical Research (CNRST) of Morocco (Smart DLSP Project - AL KHAWARIZMI IA-PROGRAM).

References

1. Vithana, K., Jayasekera, R., Choudhry, T., Baruch, Y.: Human capital resource as cost or investment: a market-based analysis. Int. J. Hum. Resour. Manag. **34**(6), 1213-1245 (2023). https://doi.org/10.1080/09585192.2021.1986106
2. Sinambela, E.A., Darmawan, D., Mendrika, V.: Effectiveness of efforts to establish quality human resources in the organization. J. Mark. Bus. Res. (MARK) **2**(1), 47–58 (2022)
3. Yongzhi, W., Kenikasahmanworakhun, P.: Antecedents Factors Affecting Employee Retention. Case Study of Dayun Group Co., Ltd. (2023)
4. Zhavoronkova, G., Zhavoronkov, V., Zavalko, K.: Management of intellectual personnel in the era of Industry 4.0. Industry 4.0 **8**(2), 68–71 (2023)
5. Ifeoma, I.-O.M., Chukwuebuka, N.E., Chidinna, O.S.: Effect of extrinsic reward system on employee's productivity in deposit money banks in Nigeria. World J. Adv. Res. Rev. **18**(3), 458–469 (2023). https://doi.org/10.30574/wjarr.2023.18.3.1092

6. Ifeyinwa, L.: Effect of reward system on the performance of sachet waters companies in Anambra State, Nigeria (2020)
7. Kushwaha, B.P., Shiva, A., Tyagi, V.: How investors' financial well-being influences enterprises and individual's psychological fitness? Moderating role of experience under uncertainty. Sustainability **15**(2), 1699 (2023). https://doi.org/10.3390/su15021699
8. Nepal, S., Mirjafari, S., Martinez, G.J., Audia, P., Striegel, A., Campbell, A.T.: Detecting job promotion in information workers using mobile sensing. In: Proceedings of the ACM on Interactive, Mobile, Wearable and Ubiquitous Technologies, vol. 4, no. 3, pp. 1–28, September 2020. https://doi.org/10.1145/3414118
9. Ghanimati, G., Darvishi, Z.A., Yazdani, H., Al-Islami, N.S.: A counterfactual thinking approach to human resources' perception of unfavorable aspects of organizational life: a two-stage methodology based on a systematic review and AHP method (2024)
10. Silitonga, A., Megawaty, D.A.: Decision support system feasibility for promotion using the profile matching method. J. Data Sci. Inf. Syst. **1**(2) (2023)
11. Liu, J., Wang, T., Li, J., Huang, J., Yao, F., He, R.: A data-driven analysis of employee promotion: the role of the position of organization. IEEE International Conference on Systems, Man and Cybernetics SMC (2019)
12. Long, Y., Liu, J., Fang, M., Wang, T., Jiang, W.: Prediction of employee promotion based on personal basic features and post features. In: Proceedings of the International Conference on Data Processing and Applications, Guangdong China, pp. 5–10. ACM, May 2018. https://doi.org/10.1145/3224207.3224210
13. bin Shafie, S., Ooi, S.P., Khaw, K.W.: Prediction of employee promotion using hybrid sampling method with machine learning architecture. Malays. J. Comput. **8**(1) (2023)
14. Pampouktsi, P., et al.: Techniques of applied machine learning being utilized for the purpose of selecting and placing human resources within the public sector. J. Inf. Syst. Explor. Res. **1**(1), 1–16 (2022). https://doi.org/10.52465/joiser.v1i1.91
15. Zulfikar, W.B., Jumadi, Prasetyo, P.K., Ramdhani, M.A.: Implementation of Mamdani fuzzy method in employee promotion system. IOP Conf. Ser. Mater. Sci. Eng. **288**, 012147 (2018). https://doi.org/10.1088/1757-899X/288/1/012147
16. Saaty, R.W.: The analytic hierarchy process-what it is and how it is used. Math. Model., 161–176 (1987)
17. Chrit, S., Azmani, A., Azmani, M.: Combining AHP and Topsis to select eligible social and solidarity economy actors for a call for grants. Int. J. Adv. Comput. Sci. Appl. **13**(11) (2022). https://doi.org/10.14569/IJACSA.2022.0131148
18. Benallou, I., Azmani, A., Azmani, M.: A combined AHP-TOPSIS model for the evaluation and selection of truck drivers. J. Theor. Appl. Inf. Technol. **101**(7) (2023)
19. Piton, G., Philippe, F., Tacnet, J.-M., Gourhand, A.: Aide à la décision par l'application de la méthode AHP (Analytic Hierarchy Process) à l'analyse multicritère des stratégies d'aménagement du Grand Büech à la Faurie. Sci. Eaux Territ. **26**(2), 54–57 (2018). https://doi.org/10.3917/set.026.0054
20. Haji, M., Kerbache, L., Al-Ansari, T.: Development of risk management mitigation plans for the infant formula milk supply chain using an AHP model. Appl. Sci. **13**(13), 7686 (2023). https://doi.org/10.3390/app13137686
21. Dira, A.F., Noor, G.P., Bangun, M.F.A., Winardi, M.A., Kamal, F., Utomo, K.P.: The role of employee engagement in green HRM to create sustainable humanist performance. EKOMBIS Rev. J. Ilm. Ekon. Dan Bisnis **12**(1) (2024)
22. Fallucchi, F., Coladangelo, M., Giuliano, R., William De Luca, E.: Predicting employee attrition using machine learning techniques. Computers **9**(4), 86 (2020). https://doi.org/10.3390/computers9040086

23. Selvi, A.J.A., Aiswarya, B.: Examining the relationship between emotional intelligence and work engagement of automobile sector employees in Chennai. Rajagiri Manag. J. **17**(2), 156–169 (2023). https://doi.org/10.1108/RAMJ-03-2022-0052
24. Hidayat, A., Syarifudin, E., Firdaos, R.: Work ability and organizational climate's effects on employees' performance (2024)
25. Nguyen Hoang, T., Dinh, K., Phuoc Minh, H., Tran Thanh, T., Le Doan Minh, D.: Opportunities and challenges for quality of human resource in public sector of Vietnams logistics industry. Int. J. Public Sect. Perform. Manag. **1**(1), 1 (2023). https://doi.org/10.1504/IJPSPM.2025.10056383
26. Aditya, M.: Pengaruh key performance index terhadap motivasi Dan Kinerja Karyawan (2024)
27. Trzaskalik, T., Wachowicz, T. (eds.): Multiple Criteria Decision Making '10–11. Scientific Publications/The University of Economics in Katowice. Publisher of The University of Economics, Katowice (2011)
28. Kumar, M., Choubey, V.K.: Sustainable performance assessment towards sustainable consumption and production: evidence from the Indian dairy industry. Sustainability **15**(15), 11555 (2023). https://doi.org/10.3390/su151511555
29. Huang, X., Yang, F., Zheng, J., Feng, C., Zhang, L.: Personalized human resource management via HR analytics and artificial intelligence: theory and implications. Asia Pac. Manag. Rev. **28**(4), 598–610 (2023). https://doi.org/10.1016/j.apmrv.2023.04.004
30. Widiastuti, T., Zulaikha, S., Cahyono, E.F., Mawardi, I., Al Mustofa, M.U.: Optimizing the intermediary function of Zakat institution using analytical network process benefit opportunity cost risk (ANP BOCR) approach. In: 4th International Conference on Islamic Economics, Business, Philanthropy, and PhD Colloquium (ICIEBP), vol. 232 (2022)
31. Hapsari, M.I., Zusak, M.B.F. (eds.) Advances in Economics, Business and Management Research, vol. 232, pp. 3–23. Atlantis Press International BV, Dordrecht (2023). https://doi.org/10.2991/978-94-6463-176-0_2

Towards Explainable Models: Explaining Black-Box Models

Bajja Nisrine[1,2,3(✉)], Tabaa Mohamed[2], and Dufrenois Franck[3]

[1] Complex Cyber Physical Systems Laboratory, ENSAM, Hassan II University, Casablanca, Morocco
`bajjanisrine@gmail.com`
[2] Pluridisciplinary Laboratory of Research and Innovation (LPRI), EMSI Casablanca, Casablanca, Morocco
`m.tabaa@emsi.ma`
[3] LISIC, 50 rue F. Buisson, ULCO Calais, Pas-de-Calais, France
`franck.dufrenois@univ-littoral.fr`

Abstract. Over the past decade, artificial intelligence has been at the heart of developing most applications that use large amounts of data "big data" collected by computing devices. The goal of this rapidly expanding field of research, which combines mathematics and computer science, is to create machines capable of simulating human intelligence. However, due to their lack of interpretability and explicability, many AI models are considered black boxes. This paper aims to review the literature of models that have been used to open up black-box models in the field of AI, thereby contributing to their explicability and making them more understandable to different types of users. In this paper we present a literature review of the models used to open the black-box models and the need for explainability. We present current techniques and models for interpreting AI models. We discuss the different approaches of explainability, including post-hoc (explainable models) and ante-hoc (interpretable models). We also cover methods such as rule extraction and model distillation.

Keywords: eXplainable Artificial Intelligence · Deep Learning · interpretability · Explainability · Rule Extraction · Model Distillation · FAT

1 Introduction

In recent years, numerous researchers have emphasized the necessity for explanatory power to dissolve the black-box nature of AI models. Many decision-making systems are considered black-boxes due to their more sophisticated and complex process in generating predictions. This inherent complexity of most AI models contributes to the opaqueness of the process leading to their predictions. While acknowledging the high performance of their models to generate accurate outputs, it becomes crucial to furnish explanations, as The European Union published in 2018, the GPDR (General Data Protection Regulation) which points to the right of citizens to receive suitable explanations about how these systems reach their decisions. Despite these challenges, strides have been taken to devise methods for interpreting and diagnosing black-box models. These methods involve using surrogate models or increasing transparency during the model-building phase. Since early 1994s, researchers have carefully sought to comprehend the nature of AI models by delving into their architecture and inner mechanisms. This survey critically examines both current and recent methods dedicated to unveiling the opacity of AI models. Some researchers elucidate the model's overall architecture, while others concentrate on explaining the model's output for specific predictions. The paper delves into the significance of interpretability during the construction of AI models. AI is frequently perceived as a black box, where data is inputted and outputs are generated, leaving many questions unanswered (See Fig. 1). However, this approach is insufficient. The solutions and answers provided by AI must be explainable in human terms. Without this capability, push back is likely to ensue.

Over the last few years, DL models have revolutionized in various fields due to their high performance. From image classification, sentiment analysis, speech understanding to natural language processing. These networks are composed of multiple layers with neurons interconnected by weights and are designed to be trained on a vast amount of data [1]. Their capability to detect patterns during the training phase has significantly raised their use in face and speech recognition. Due to their highly efficient predictions, people consider how comfortably they can trust and rely on their outputs, particularly in overly sensitive domains [2]. The primary challenges in interpreting these models are cited by Hassija et al. (2023) [2], beginning with understanding the optimization process of the learned weights, followed by the non-linear aspect that can arise from a small variation in the input to a huge change in the output, the third challenge is the enormous number of parameters used in DL models. The need for XAI is a turning point. The initial concepts for providing explainability emerged in 2015, DARPA's initial programme resulted in the development of modified machine learning techniques that produce models that are explainable. This enables end users to understand and trust AI systems. In 2017, their XAI research began, consisting of 11 research teams, and ends in 2021 with the creation of a central, publicly accessible XAI toolkit.

In this survey we propose to:

(i) discuss the black-box nature of AI models,
(ii) the need for interpretability and explainability in AI models, with some
(iii) use cases applications of XAI in AI models. By exploring the state-of the-art techniques used to explain models, we propose a well-informed exploration of the methodologies employed. At the end of this paper, we will discuss the
(iv) challenges of integrating explainability in ML models.

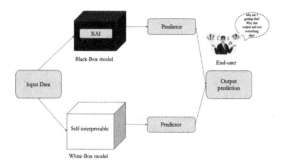

Fig. 1. Black-box and White-box model

2 Background

2.1 Explainability

Most active academic research on XAI lacks clear and definitive definitions for the term. A precise and satisfactory explanation has not yet been proposed, resulting in ambiguity and a demand for more conceptual clarifications in the field of AI. Broniatowski (2021) [3] encompasses the need for both interpretation and explanation to enhance explainability. Much research has been carried out to either use a simplified model or generate explanations to understand the inner mechanism of the model. Mittelstadt (2018) [4] distinguishes between the use of simplified models and explanations, highlighting the philosophical and sociological differences between them. Simplified models allow end users to answer "what if" questions or generate contrasting explanations to clarify the model. Explanations are described as the degree of completeness by which the entire causal chain of an event can be explained while providing a comprehensive understanding, rather than using a simplified model. In other terms, explainability can refer to how well a model can provide insights that are both accurate and understandable. These perspectives collectively underline the necessity for AI models to provide clarified insights, tailored explanations and human-centric approach to build understanding and trust.

"If you can't explain it simply, you don't understand it well enough."
– Albert Einstein

2.2 Interpretability

Interpretability occurs when accountability, trust and fairness are involved in deploying a particular decision-making system. It plays a vital role in ensuring the trust of end users and stakeholders. It is preferable to know the reason for a decision rather than just accepting it without an answer for our questions. By ensuring trust and identifying biases for ongoing model development, this helps to make AI models trustworthy by stakeholders.

Ennab (2022) [5] highlights the importance of interpretability in healthcare AI models to ensure accountability, trust, and compliance with data protection regulations. Lipton (2018) highlights the challenge of defining and achieving interpretability in machine learning. Broniatowski (2021) [3]distinguishes between interpretation and explanation, emphasizing the need to customize system output to different user preferences. Framling (2020) [6] proposes Contextual Importance and Utility (CIU) as a model-agnostic approach to generating human-like explanations of black-box outcomes. These studies emphasize the importance of interpretability in AI models to ensure transparency, trust, and user understanding. However, achieving interpretability can be complex and challenging. Due to the increasing use of artificial intelligence-based systems in recent years interpretability has become an essential way for AI models to ensure their development in an ethical and responsible way.

"Interpretability is the degree to which a human can understand the cause of a decision."

– Tim Miller

"Interpretability is the degree to which a human can consistently predict the model's result."

– Been Kim

2.3 Explainable and Interpretable Models

Before delving into the techniques used for elucidating the inner workings of black-box models, we should first make a crucial distinction between interpretable models and those considered explainable. Interpretability refers to the extent to which user can understand the relationship between the input features and the output predictions answering the question "how the model arrives to its predictions?". On the other hand, explainability aims to provide justifications and clarifications describing "why a particular input gives a specific output?". Explainable models are considered as interpretable and not vice versa [9].

Through a thorough exploration of numerous research studies, it has been determined that explainable models provide post-hoc explanations, while interpretable models offer ante-hoc explanations. Lisboa et al. (2023) [14] have proposed that for a robust explanation and interpretation, intelligibility (immediately understandable), faithfulness, stability, sparsity, transparency, comprehensibility and model inspection (model representations) should be taken in account. Figure 2 depicts the trade-off between interpretability and accuracy. The higher the accuracy, the lower the interpretability. Linear regression models are easily understood and interpreted. While Support Vector Machines or Random Forest are considered not easily understood, there is no white box notion here and we came to talk about gray-box. On the other hand, Neural Network with high number of parameters become extremely difficult to understand their representations. Due to this, interpretability is a spectrum to unveil the opaqueness of these models. In the next section, we will present the founded methods and techniques for both interpretable and explainable models.

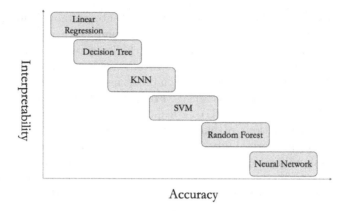

Fig. 2. Trade-off between Interpretability and accuracy.

3 Exloring the Inner Mechanisms: Open the Black-Box Model

In this section, we will discuss several state-of-the-art techniques for interpreting and explaining ML models. To contextualize the research, it is essential to understand the distinction between explainable and interpretable models. Interpretable models aim to identify the cause and effect by associating each cause (input) to its effect (output)[1].

Interpretability seeks to understand the whole decision-making process behind all predictions. While explainable models underscore the clarification and justification of those outputs answering the why question. Adadi et al. (2018) [17] define explanation as the way to justify, clarify, control, improve and discover. However, Burkart and Huber (2020) [16] claims that an explanation should be contrastive, concise, selected, and social.

The literature reveals diverse perspectives on the techniques used for explainability of ML models. They have found that explanations should be unprecise, unstable, and lacking in insight into their correctness and reliability [18].

4 Interpretability Approaches

Interpretability plays a keystone in AI models. Interpretable models provide explanations through their understandable architecture and design. It's an approach to constrain the interactions between the input features and the output in a comprehensible way. Interpretable models are defined by Lisboa (2023) [14] as ante-hoc explanations. They seek to elucidate the model during the construction phase before its deployment. Interpretable models can be defined as inherently transparent models, exemplified by rule-based models, or as models whose structures are interpretable, as seen in decision trees. An important consideration in

---

[1] https://www.bmc.com/blogs/machine-learning-interpretability-vs-explainability/.

the use of interpretable models involves the model-based dimension, which comprises two distinct types: (1) specific and (2) agnostic. (1) used for a certain model and not applicable to all models, conversely, (2) is used for any type of model. Additionally, scope-based dimension emphasizes the scope of interpretability, local aim to provide explanation for a specific prediction (outcome), while the global scope aims to provide interpretations for the model as a whole. While the previous paragraph emphasized the understanding of the approaches of interpretable models, the following section sheds light on inherently interpretable (transparent) models. –Linear regression models: They aim to provide each feature (variable) with a specific weight, and the predicted outcome is the sum of the weights of the input features. The weights indicate the contribution of each feature in the prediction of the model.

–Decision Tree: when the relationship between features and target variable becomes nonlinear, linear models fail. On the other hand, decision trees capture the interaction among variables (nodes) and the target (leaf nodes). The construction of a decision tree relies on the utilization of splitting methods to progress from each node to the subsequent nodes, starting from the root node. Notably, Decision trees are considered interpretable due to the extraction of decision rules, such as "if a feature x_i has been greater than... then....". In addition, contrastive rules like "what if "scenarios can be derived from decision trees.

–Rule-based models: decision rules have a general and comprehensible structure "IF a condition (or a set of conditions) is(are) met THEN generate this prediction". For example, "If hours of study (X_1) are greater than 5 h and attendance in class (X_2) is greater than or equal to 80.

–M-of-N rules: they specify that a certain number M of conditions out of a set of conditions N must be satisfied for the rule to be activated. They allow for a more flexible decision-making process by considering multiple conditions in a rule.

–Naives bayes models: they use the Bayes theorem to calculate the probability of a class based on the value of the feature. By using this process, we can interpret the contributions of each feature to the prediction of a particular class.

–K-Nearest Neighbors: they compute the nearest neighbors using a similarity measure to calculate the distances between instances. When we want to explain a feature, we can retrieve the k neighbors that were used for the prediction.

–GAM: General Additive Models are based on a set of simple functions that can be easily understood and interpreted. The model is a sum of functions of each input feature, and the coefficients of the functions can be easily interpreted to see how each feature affects the output.

5 Explainability Approaches

The rise of the landscape of artificial intelligence has prompted a demand in the necessity for understanding the why behind the decisions of AI models. Before delving into the methods extracted from various state-of-the-art literature reviews, it's essential to establish a clear definition of the concept of explanation, and what is needed to provide a good explanation. Explainability provides answers to 'why' questions, aiming to render the output of the ML model understandable to human beings. Simultaneously, according to Gilpin (2018) [9], an explanation is defined as a summary of the reasons behind a ML model's decision-making process, offering insights into the 'why' and 'why should (not)' aspects. Meanwhile, Burkrat (2020) [16] emphasizes the necessity for a good explanation and suggests that, for a comprehensive explanation, it is crucial to address the question of 'What to explain?' (output or the model), 'Whom to explain to?' (target end-user), and 'How to explain?' (method used).

While accuracy remains an essential factor for evaluating model performance, it is imperative to strike a balance between accuracy and interpretability. This section delves into post-hoc explanations as a means of maintaining both accuracy and transparency.

5.1 Post-Hoc Model Interpretability

Seeking to improve the interpretability of ML models, post-hoc explanation methods have emerged as invaluable tools. Among these, visualization techniques and example-based techniques.

Visualization Techniques. The human visual system serves as the most efficient channel connecting with the human brain. Numerous studies underscore the crucial role that visualization plays in bridging the gap between opaque model predictions and human interpretability. Tamagnini (2017) [15] supports this with the introduction of Rivelo, a graphical analytics interface that helps analysts to better understand the reasons behind the predictions of binary classifiers. There exists a diverse array of techniques within the realm of visualization interpretability methods.

–Heatmaps: Heatmaps provide visual representations that highlight the importance or relevance of different features or areas within the input data. Taking the example of an image in the interpretation of a neural network, heatmaps can visualize the intensity of colors that corresponds to the level of influence allowing users to identify the critical features that influence the model's decision (Figs. 3 and 4).

Visualization techniques are based on the contribution of each feature in the model's prediction, by affecting to each feature a relevance score that indicates the degree of its contribution. From the most used techniques to calculate the relevance score, Sensitivity Analysis (SA) using the formulation (1). SA evaluates the sensitivity (uncertainty) of the input data to the outcome by evaluating the gradient. The most relevant features are those to which the output is most sensitive [12]. The resulting vector of SA will contain the relevant features with a Ri positive and others with a Ri set to zero as irrelevant features. By drawing the heatmap, we can distinguish what pixels of an image are relevant for a classification. On the other hand, Simple Taylor Decomposition aim to provide an explanation of the question why these pixels are relevant?' One can also use equation arrays:

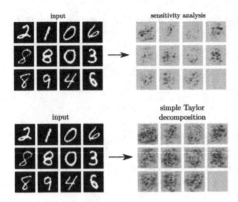

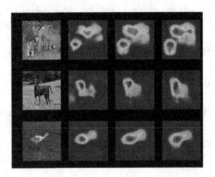

Fig. 3. sensitivity analysis and simple Taylor Decomposition applied to Convolutional DNN, heatmaps (as explanations). Red and Blue colors indicate the positive and negative relevance score. This figure is extracted from [12] (Color figure online).

Fig. 4. Example of Saliency Maps.

$$R_i = \left\| \frac{\partial}{\partial x_i} f(x) \right\| \tag{1}$$

Example-Based Techniques. They are model agnostic, because they can make any model interpretable by generating examples. These techniques rely on a specific instance to provide insights into the decision-making process of the model. Sheikh Rabiul (2021) [10] describes the concept of example-based explanations as "X is similar to Y and Y caused Z, so the prediction says X will cause Z". The following are some Example-based techniques:

–Prototypes: also nominated archetype or artifact. It's the selection of a set of representative instances that characterizes a class. It's considered as an example because the user can understand the model reasoning by looking at instances similar to the prototype.

–Adversarial Examples: are examples and records used to manipulate the input data (making small changes and perturbations). Adversarial attacks involve generating perturbations to the input data to deceive the model and cause misclassification. Researchers can gain insights into the vulnerabilities and decision boundaries of the model.

–Counterfactuals: These explanations aim to provide alternative scenarios by identifying minimal edits to the input features that would lead to a different model output. In the case of credit approval, they reveal the changes a client should make to secure credit approval. Counterfactual explanations are easily understood by humans because individuals tend to comprehend the impact of an event by contrasting it with alternative events. Answer the question (what I must change to get the desired output?

6 Interpretability Methods

6.1 Knowlege Extraction

Knowledge extraction refers to methods used to extract meaningful insights from the model. These methods aim to uncover hidden patterns within the relationship between features. Two distinct methods will be explored in this section: Rule extraction and model distillation.

Rule Extraction. Rule extraction is the process of converting a black-box model into a human-understandable rules that describe its decision-making process. The extracted rules are in the form of IF-THEN statements providing a comprehensive representation of the behavior of the model. Kenny (2021) [19] explored the effect of visual explanations and error rates and proposed a method for extracting feature weights from convolutional neural network using MNIST dataset. Rule extractions are divided into three types: Decompositional, pedagogical and eclectic rules.

Model Distillation. Model distillation, also known as knowledge distillation, is the process of extracting insights from a sophisticated (teacher) model (i.e. DNN) to a simplified (student) model in order to maintain the accuracy of the black-box model. This concept is based on training the student model on a smaller dataset using the output of the teacher model as labels.

6.2 Feature Knowledge

The features (variables) of the model are the basis for its predictions, forming the fundamental pillars that underline the decision-making process. Understanding the contribution of each feature in the prediction is essential to explain the model's process. Understanding the contribution of individual features is insufficient; therefore, it is imperative to delve into the interactions between features to gain a thorough understanding of the model's decision-making process.

Feature Importance: Feature importance methods, also called univariate methods, are based on identifying the most influential features for the prediction by understanding how individual features influence the prediction outcomes. The method is based on retraining the model without a specific instance and trying to identify if the predictions largely vary, indicating the importance of an instance [13].

Feature Interaction: Feature interaction is also called bivariate methods; they aim to identify the interaction between features by selecting the most influential ones. These methods go beyond analyzing the impact of individual features by exploring the effect between more features. They help in understanding patterns between data and gain insights into how multiple features affect the model's output.

LRP (Layer-Wise Relevance Propagation). LRP is a method for explaining black-box models, based on attributing relevance score to each input feature. Starting from the output layer, LRP uses backpropagation back to the input layer to measure the contribution of each feature to the model's prediction. The activation is calculated in each layer, and the score is calculated based on the activation.

SHAP (SHapley Additive ExPlanations) (feature Importance and Feature Interaction). SHAP is a technique to interpret black-box model based on game theory. The concept is presented by Llyod Shapley in 1953, when he seeks to know the contribution of each player in a coalition game. Then researchers adopt this idea in ML models in order to explain black-box models, the "game" can be defined as the instance to be explained, "players" as the instance's features and the "gain" is the difference between the prediction of the instance and the average of the predictions of all instances. SHapley values underscore the contribution of each feature in the final prediction. The term "additive" emphasizes the additive nature of shapley values. SHAP is considered as a powerful technique for both local and global explainability. SHAP specify the explanation as: The base value (0.5013) and the predicted value is 0.71, the visual presents the effect of each feature in decreasing (feature value with blue) or increasing (feature value with pink) the predicted value. Ball Possession has a significant effect in decreasing the prediction, while Goal Scored increases the prediction. (See Fig. 5)[2].

Fig. 5. SHAP (SHapley Additive exPlanations) visualization of the effect of each feature in the prediction of the model.

DeepLIFT. DeepLift was proposed by Shrikumar et al. (2017) [8], it is a Method to interpret Deep Learning models by assigning contribution score to

---

[2] https://www.kaggle.com/code/dansbecker/shap-values.

each input feature of a specific output. DeepLift uses backpropagation to calculate the difference between the activation of the actual neuron and a reference[3] base input. This method underscores the changes in the output according to the input features, contrastively to LRP that emphasizes the understanding of neurons and activations at each layer of the network.

Anchors. Anchors were presented by the inventor of LIME Ribeiro [7], it is a method to explain black-box models using decisions rules. When tasked with explaining a given instance, the approach involves locally generating perturbed predictions. Over a series of iterations, each iteration produces a candidate rule with a new feature. The process continues until a rule with the highest estimated precision is identified (Table 1).

Table 1. The difference between LIME and Anchors.

LIME	– Perturbating instances around a data point to be explained. – Training an interpretable model on perturbated data that approximate the black-box model locally. – Model-based.
Anchors	– Rule-extraction based approach. – Generate candidate rule. – Select the rule with the highest estimated prediction.

LIME. LIME is an interpretable technique used to explain the predictions of a black box model by learning locally an interpretable model around a sample. LIME generates a new dataset based on altering the original input locally with slight variations. These new data points are then used to train a proxy (LASSO, LR, DT) model. This surrogate model should approximate the black-box model around the selected sample. LIME is a model-agnostic because it can be applied to any type of black box model regardless of its architecture [7]. The formulation of LIME denoted as:

$$\psi(x) = \arg\min_{g \in G} \mathcal{L}(f, g, \pi x) + \Omega(g) \quad (2)$$

$$\mathcal{L}(f, g, \pi x) = \sum_{z, z' \in Z} \pi_x(z)(f(z) - g(z'))^2 \quad (3)$$

$\psi(x)$ represents the explanation, $\mathcal{L}(f, g, \pi x)$ is the loss function which represents the dissimilarity between the predicted values of the black-box model f and the interpretable model g. π_x presents the proximity to the original instance x, and

---
[3] The reference activation in DeepLIFT is determined by choosing a reference input and propagating activations through the network.

$\Omega(g)$ is the regularization term applied to the model g. $\mathcal{G}$ is the set of interpretable models. The optimization problem seeks to find the interpretable model g from $\mathcal{G}$ that minimizes the loss function. The goal is to find an interpretable model that is an accurate approximation of the model f. LIME provides a visual explanation as illustrated in (see Fig. 6)[4]. The figure represents the relationship between features (RM, LSTAT, PTRATIO, NOX, DIS, AGE) and predicted value, showcasing the contribution of each feature in the outcome of the prediction. LSTAT, NOX have the highest positive score while RM have a negative influence on the model's prediction.

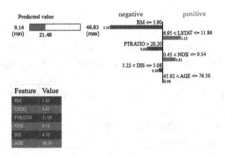

Fig. 6. LIME explanation using Boston housing dataset.

7 Evaluating the Quality of Explanations

Many methods focused on providing explanations to make AI models more interpretable, but what really makes a good explanation or what are the criteria that should be taken to evaluate the quality of an explanation to be understandable by humans? Kim (2021) [11] propose a methodology based on three categories:

–Human-grounded: this methodology is based on human experiment to evaluate the quality and the comprehensibility of the explanation rather than its ultimate goal of providing an explanation of the prediction or model.

–Application-grounded: is based on human expertise knowledge in the domain to evaluate the accuracy and the application in real world of the explanation.

–Functionality-grounded:
Application-grounded and human-grounded approaches rely on human experiments. Functionality-grounded proposes an alternative using formal descriptions as proxies. The reason for this is that human experiments are time consuming and expensive for machine learning models. However, determining and identifying the appropriate proxies remains a challenge for functionality-grounded approaches.

---

[4] https://www.kaggle.com/code/dansbecker/shap-values.

8 Challenges and Trade-Offs

Explainability of models is essential in areas where a detailed explanation of predictions is paramount; in domains such as healthcare, and specifically in diagnostic imaging, for example, if an AI model can diagnose cancer based on images, specialists can't have full confidence in that prediction, which has an impact on patients if they don't have a clear explanation. In contrast, in applications that sometimes require more powerful models than interpretable ones, explainability is an add-on.

9 Conclusion

In recent years, XAI has been considered as a field of interdisciplinarity that seeks to make black-box models interpretable. In this paper, we conducted a comprehensive review of existing methods used in the field of explainability, Our aim is to provide an overview of the current state of the art and to highlight possible directions for future research in the field. While acknowledging the existing accuracy-interpretability trade-off, it is imperative to leverage explainable tools to ensure the interpretability of models, especially within domains considered sensitive. Due to the transparency of models, users can detect biases and anomalies, thereby enhancing their durability. In the future, our aim is to combine accuracy and transparency, ensuring both robust predictive results performance and comprehensibility.

References

1. Lokhande, P., Surati, S., Trivedi, H., Shrimali, B.: Deep learning model based face mask detection for automated mandation. In: International Conference on Advancement in Computation & Computer Technologies (InCACCT) 2023, pp. 166–170 (2023)
2. Hassija, V., Chamola, V., Mahapatra, A., Singal, A., Goel, D., Huang, K., Scardapane, S., Spinelli, I., Mahmud, M., Hussain, A.: Interpreting black-box models: a review on explainable artificial intelligence. Cogn. Comput. **16**, 45–74 (2023)
3. Broniatowski, D.A.: Psychological Foundations of Explainability and Interpretability in Artificial Intelligence (2021)
4. Mittelstadt, B.D., Russell, C., Wachter, S.: Explaining explanations in AI. In: Proceedings of the Conference on Fairness, Accountability, and Transparency (2018)
5. Ennab, M., Mcheick, H.: Designing an Interpretability-Based Model to Explain the Artificial Intelligence Algorithms in Healthcare. Diagnostics, 12 (2022)
6. Främling, K.: Decision theory meets explainable AI. Explainable, Transparent Autonomous Agents Multi-Agent Syst. **12175**, 57–74 (2020)
7. Ribeiro, M., Singh, S., Guestrin, C.: "Why Should I Trust You?": explaining the predictions of any classifier. In: Proceedings of the 22nd ACM SIGKDD International Conference on Knowledge Discovery and Data Mining (2016)
8. Shrikumar, A., Greenside, P., Kundaje, A.: Learning Important Features Through Propagating Activation Differences. Presented at the (2017)

9. Gilpin, L.H., Bau, D., Yuan, B.Z., Bajwa, A., Specter, M.A., Kagal, L.: Explaining explanations: an overview of interpretability of machine learning. In: 2018 IEEE 5th International Conference on Data Science and Advanced Analytics (DSAA), pp. 80–89 (2018)
10. Rabiul Islam, S., Eberle, W., Khaled Ghafoor, S., Ahmed, M.: Explainable Artificial Intelligence Approaches: A Survey.arXiv e-prints, arXiv-2101 (2021)
11. Kim, B., Doshi-Velez, F.: Machine learning techniques for accountability. AI Mag. **42**, 47–52 (2021)
12. Montavon, G., Samek, W., Müller, K.: Methods for interpreting and understanding deep neural networks. Digit. Signal Process. **73**, 1–15 (2017)
13. Ali, S., Abuhmed, T., El-sappagh, S., et al.: Explainable Artificial Intelligence (XAI): What we know and what is left to attain Trustworthy Artificial Intelligence. Information Fusion, vol. 99, p. 101805 2023
14. Lisboa, P.J., Saralajew, S., Vellido, A., Villmann, T.: The coming of age of interpretable and explainable machine learning models. Neurocomputing **535**, 25–39 (2023)
15. Tamagnini, P., Krause, J., Dasgupta, A., Bertini, E.: Interpreting black-box classifiers using instance-level visual explanations. In: Proceedings of the 2nd Workshop on Human-In-the-Loop Data Analytics (2017)
16. Burkart, N., Huber, M.F.: A Survey on the Explainability of Supervised Machine Learning. *ArXiv*, abs/2011.07876 (2020)
17. Adadi, A., & Berrada, M. (2018). Peeking Inside the Black-Box: A Survey on Explainable Artificial Intelligence (XAI).*IEEE Access, 6, 52138-52160.*
18. Slack, D., Hilgard, S., Singh, S., Lakkaraju, H.: Reliable Post hoc Explanations: Modeling Uncertainty in Explainability. Neural Information Processing Systems (2020)
19. Kenny, E.M., Ford, C., Quinn, M.S., Keane, M.: Explaining black-box classifiers using post-hoc explanations-by-example: The effect of explanations and error-rates in XAI user studies. Artif. Intell. **294**, 103459 (2021)
20. Li, Y., Wang, H., Dang, L.M., Nguyen, T.N., Han, D., Lee, A., Jang, I., Moon, H.: A deep learning-based hybrid framework for object detection and recognition in autonomous driving. IEEE Access **8**, 194228–194239 (2020)

Advanced NLP and N-Gram Techniques in Financial News Sentiment Analysis: Diverse Machine Learning Approaches

Oussama Ndama[(✉)] and El Mokhtar En-Naimi

DSAI2S Research Team, C3S Laboratory, FST of Tangier, Abdelmalek Essaâdi University, Tetouan, Morocco
oussama.ndama@etu.uae.ac.ma, en-naimi@uae.ac.ma

Abstract. This study explores the complex field of financial news sentiment analysis, which is influenced by various elements such as business-specific or political news, user opinions, and the ever-changing structure of regional financial markets. Positive news has the capacity to stimulate growth in markets, but negative news might trigger falls. This study focuses on developing a strong financial sentiment analysis model that is specifically designed to identify feelings inside financial news headlines, in light of these complexity. This research introduces a novel methodology that gives a revolutionary paradigm for analyzing sentiment in financial news. It leverages advanced natural language processing techniques including N-Gram analysis, Term Frequency-Inverse Document Frequency (TF-IDF), and a variety of machine learning methodologies. The study rigorously compares the Uni-Gram, Bi-Gram, and Tri-Gram approaches using five independent machine learning algorithms. Empirical evaluations highlight the effectiveness of preprocessing, polarity analysis, Uni-gram, and TF-IDF as main methods for extracting features. When combined with Linear SVC as the classifier, this method obtains a remarkable accuracy rate of 94%. This paper not only recognizes the complex nature of financial sentiment research but also outlines a promising methodology that skillfully addresses these complexities, offering deep insights into sentiment classification in financial news.

Keywords: Financial News Sentiment Analysis · Natural Language Processing · N-Gram Analysis · Multi-Class Classification · Machine Learning Classifiers

1 Introduction

Within the field of financial news sentiment analysis, our study addresses the intricate nature of multi-class classification [1], which is made more difficult by the various factors that shape market emotions. The statistics used in this study were carefully obtained from reliable sources such as Consumer News and Business Channel (CNBC), the Guardian Business, and Reuters. They cover a wide range of news related to the U.S. economy and stock market from 2017 to 2020.

Traversing various datasets poses a difficult terrain marked by diverse formats, frequencies of updates, and availability of content. Nevertheless, this variety also enhances our study, allowing for a more comprehensive comprehension of the emotional changes within the financial news domain.

The issue of multi-class classification in this context involves two main aspects: firstly, accurately detecting sentiments, whether they are positive, negative, or neutral, within these multiple datasets; and secondly, integrating these different sources into a coherent framework for analysis. When numerous data sources are used, it is important to have a strong approach to deal with the challenges that arise from differences in text structures, availability of information, and time dimensions [2].

Our work utilizes a wide range of advanced Natural Language Processing (NLP) approaches, such as n-gram analysis and TF-IDF, along with a set of complex machine learning algorithms [3]. The techniques, specifically Linear Support Vector Classifier (LinearSVC), Logistic Regression, Multinomial Naive Bayes, Random Forest, and Gradient Boosting, are essential in our efforts to extract meaningful sentiments from this large and heterogeneous dataset.

The combination of many data sources and the complexities of multi-class classification present both a difficulty and a chance for advancement. This statement highlights the intricate nature of financial sentiment analysis and the potential for groundbreaking advancements. It presents an opportunity to develop a pioneering approach that can effectively analyze the vast amount of data available and extract valuable insights necessary for comprehending and forecasting market fluctuations.

2 Related Studies

Previous research in the field of financial news sentiment analysis has explored different areas, including strategies for extracting sentiment and the application of machine learning algorithms for sentiment classification. The purpose of this part is to analyze and combine the many methods, approaches, and results from prior research efforts, offering useful insights and expanding on the groundwork established by earlier studies in this field.

B. Meyer, M. Bikdash and X. Dai's study [4] probes fine-grained financial news sentiment analysis, spotlighting the influence of constant news flows on market sentiments. Their innovative methods, utilizing Natural Language Processing, reveal shortcomings in existing sentiment analysis. Comparing lexicon-based and machine learning approaches, the research consistently favors machine learning techniques. These findings equip financial practitioners with potent tools to integrate nuanced news sentiment into risk and pricing models.

K. Mishev, A. Gjorgjevikj, I. Vodenska, L. T. Chitkushev, and D. Trajanov's research on sentiment analysis in finance explores various methodologies' effectiveness in extracting sentiments from financial news. They discover that utilizing contextual embeddings, particularly NLP transformers, outperforms lexicons and fixed encoders in sentiment analysis, even in scenarios with limited data availability. Furthermore, their study reveals that distilled NLP transformer models exhibit similar performance to larger models, indicating their suitability for practical deployment in production environments [5].

Kouissi et al. [6] propose a hybrid approach for Twitter sentiment analysis, blending Dynamic Case-Based Reasoning, Multinomial Logistic Regression, and a Multi-Agent System. Their method classifies tweets into positive, negative, or neutral sentiments on various topics. It involves preprocessing, TF-IDF content similarity scoring, Multinomial Logistic Regression for classification, and K-Nearest Neighbors for finding similar tweets. This adaptable method focuses specifically on analyzing Covid-19-related tweets to predict user behaviors and sentiments during critical situations.

X. Man, T. Luo and J. Lin present a comprehensive survey on Financial Sentiment Analysis (FSA) to address the growing need for analyzing extensive financial text data in the realm of NLP. Their research covers diverse aspects of FSA, including data sources, lexicon-based methods, traditional machine learning, and recent advancements in deep learning like word embedding, Convolutional neural network (CNN), Recurrent neural network (RNN), Long Short-Term Memory (LSTM), and attention mechanisms. Additionally, the study outlines future directions such as large unsupervised contextual pretraining, hierarchical approaches, joint learning, transfer learning, and potential applications in FSA [7].

B. Agarwal, explores financial sentiment analysis, aiming to understand how news impacts stock markets. The study introduces a neural network model that refines general word embeddings into domain-specific ones, bolstered by a knowledge-base. Addressing challenges in NLP, particularly with rare words and shifting polarities, the model excels in SemEval-2017's financial sentiment analysis dataset. Results highlight its top-tier performance on Twitter and news headlines, proving its effectiveness in this field [8].

M. Vicari and M. Gaspari explore using deep learning, especially LSTM models, to predict market sentiment from news headlines. By analyzing 25 daily headlines related to the Dow Jones industrial average from 2008 to 2016 (extended to 2020), they aim to create an algorithmic trading strategy. The study tests these predictions across real-world scenarios to assess their effectiveness in forecasting market sentiments [9].

3 Methodology and Materials

The methodology and materials section are crucial components in establishing a strong foundation for multiclass classification in the field of sentiment analysis applied to financial news headlines. This section provides a detailed description of the dataset used, explaining the careful preprocessing approaches employed to enhance the data's suitability for sentiment analysis and the classification of financial news sentiments. The essential aspect of this methodology involves the intentional incorporation of polarity analysis, n-gram analysis, and TF-IDF methodologies. These approaches act as fundamental principles, enabling the systematic retrieval of emotion orientation, linguistic patterns, and the relative significance of terms within the textual material. Furthermore, the intentional choice and utilization of various machine learning algorithms as classification models strategically empower this framework to fully understand and effectively handle the complex subtleties involved in analyzing sentiments across multiple classes within the ever-changing field of financial news.

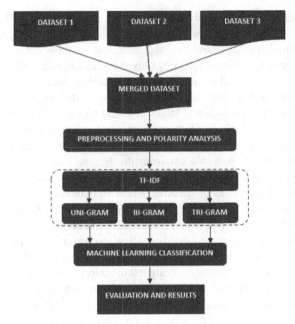

Fig. 1. Architecture of Our Financial News Sentiment Analysis Methodology.

3.1 Data Description

The dataset utilized in this study is an amalgamation of meticulously scraped headlines from CNBC, the Guardian, and Reuters official websites. These datasets collectively form a comprehensive overview of the U.S. economy and stock market, spanning a temporal scope of the past years (2017–2020). Specifically, the dataset sourced from CNBC encompasses an array of valuable information including headlines, last updated dates, and concise preview texts of articles, spanning from the conclusion of December 2017 to July 19th, 2020. Meanwhile, the dataset extracted from the Guardian Business predominantly includes headlines and last updated dates of articles, covering the same timeframe as CNBC's dataset, as the Guardian Business does not provide preview texts. Additionally, data gleaned from Reuters enriches this comprehensive dataset with headlines, last updated dates, and preview texts of articles, covering the period from the terminus of March 2018 to July 19th, 2020. This meticulous curation ensures a diverse yet encompassing dataset reflective of the dynamics and sentiments prevalent in the U.S. economy and stock market over the specified duration [10].

3.2 Text Preprocessing

The dataset preparation for our study incorporates a structured sequence of preprocessing steps to refine textual data quality for advanced analytics. Initially, we aggregate datasets from varied outlets, including CNBC, the Guardian Business, and Reuters, to form a consolidated analytical base. This amalgamation process is critical for handling diverse formats and ensuring uniformity across different data sources.

Subsequent text cleaning involves a meticulous approach to standardize the corpus: we first normalize text to lowercase to achieve uniformity. This is followed by purging punctuation, special characters, and irrelevant symbols through precise string manipulation techniques, thereby retaining only essential linguistic elements. To ensure consistency despite the varied source formats, this step includes comprehensive rules for symbol and punctuation removal applicable across all data types.

Tokenization and the exclusion of stop words and numeric characters further refine the dataset. This stage selectively filters out common English stop words and numbers, retaining significant phrases vital for sentiment analysis. Our method's reliability is bolstered by a detailed stop words list from the Natural Language Toolkit (NLTK) corpus, ensuring that no source-specific stop words are overlooked.

Lemmatization, the process of reducing words to their base form, is the final step in our preprocessing journey. Utilizing the WordNetLemmatizer from NLTK, we standardize words to their root forms, which is crucial for maintaining semantic consistency across various sources. This technique aids in aligning terminologies and concepts from different news outlets, enhancing the dataset's logical coherence and interpretability.

Our preprocessing methodology is designed to address the challenges posed by the diverse formats of financial news headlines. By implementing these systematic steps, we ensure the consistency and accuracy of the processed text, laying a robust foundation for reliable sentiment analysis in the financial domain. This enables the precise identification of sentiment trends within the collated headlines, thereby significantly contributing to the study's efficacy [11–13] (Table 1).

Table 1. Example of The Preprocessed Financial News Headlines.

Index	Headline
0	jim cramer better way invest covid19 vaccine gold rush
1	toyota turn ai startup accelerate goal robot home
2	trump torn us china trade deal official push fulfill term
3	ceo tell trump hiring american without college degree
4	china february export seen falling two-year import reuters poll

We will present the preprocessed data in order to get insight into the most frequently occurring word in the dataset. The Word-Cloud API is utilized to exhibit the outcome for the financial news dataset. Figure 1 provides a concise overview of the terms that are most frequently found in the headlines.

The most frequently used words are say, china, trade, uk, business, deal, coronavirus, bank, trump, billion, source, company, new, brexit, stock, market, economy etc. (Fig. 2).

Fig. 2. Word-Cloud for Financial News Headlines.

3.3 Polarity and Scoring Method

The research methodology relies on the complex field of sentiment analysis, a crucial tool used to identify the polarity whether positive, negative, or neutral present in textual data. The field of advanced NLP is fundamental in multiple areas, such as data mining and text analytics. It enables the extraction and examination of opinions expressed on numerous platforms, such as blog posts, comments, reviews, and tweets [14, 15].

The objective of our study is to classify financial news headlines using the VADER sentiment analysis tool, which is a module in NLTK that can accurately determine sentiment ratings based on the semantic orientation of words. VADER operates as a sentiment analyzer that follows a set of rules. It assigns sentiment scores to different parts of text by classifying phrases as positive or negative, depending on their semantic orientation [16]. VADER's primary premise is based on using a lexicon that maps lexical information to intensities of emotions. This allows VADER to calculate sentiment scores for a given text. This method computes sentiment ratings by aggregating the intensity of each word in the text, providing important insights on the overall polarity of the sentiment.

VADER provides sentiment scores that consist of probabilities for positive, neutral, and negative sentiments [17]. The total of these probability adds up to 1. In addition, the compound score is a normalized statistic that combines all lexical evaluations to create a single measure of sentiment, ranging from -1 to 1. By utilizing VADER in our study, we are able to conduct a thorough assessment of sentiment, which assists in the detection and categorization of text according to their compound scores. The use of predetermined thresholds, such as considering a compound score more than 0 as positive, a score of 0 as neutral, and a score less than 0 as negative, allows for a detailed analysis of the attitudes expressed in financial news headlines. This methodology enables us to extract significant insights from these attitudes, establishing the foundation for educated decision-making based on data.

Figure 3 presents an overview of the financial news headlines dataset's distribution across the three sentiment classes post-calculation of polarity and score.

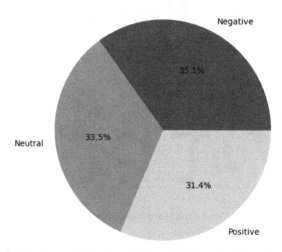

Fig. 3. Sentiment distribution of Financial News Headlines Dataset.

3.4 N-Gram Analysis

N-Gram analysis is a fundamental technique in NLP that involves examining consecutive sequences of n elements, such as words or letters, in a given text corpus [18]. N-Gram, which are sequences of words, are crucial for capturing complex linguistic patterns, enhancing language models, and enabling various tasks including text prediction and processing. Every N-Gram, regardless of whether it is a Uni-Gram, Bi-Gram, or Tri-Gram, captures the contextual meaning of the text [19]. A Uni-Gram, refers to a single word sequence such as "high", "market" or "growth", a Bi-Gram, refers to a sequence of two words, such as "high market" or "market growth", a Tri-Gram, refers to a series of three words, such as "high market growth". This methodology is widely applicable in various domains of data science and computer science, playing a significant role in tasks such as language modeling, semantic feature extraction, spelling correction, machine translation, and text mining. The classification of N-Gram, as shown in the table, emphasizes their significance in comprehending sequences of words and symbols, particularly in crucial tasks such as sentiment analysis, text classification, and machine translation. The versatility and adjustability of N-Gram models, customized for different n-values, highlight their widespread impact and usefulness in deriving significant insights from textual data in various analytical fields.

3.5 TF-IDF

TF-IDF, short for Term Frequency-Inverse Document Frequency, is a robust statistical measure used in information retrieval and machine learning to assess the importance of terms in a set of documents [20]. The TF-IDF score ((TF-IDF(t, d, D))) for a term (t) in a document (d) from the document set (D) is determined by multiplying the term frequency (TF) ((TF(t, d))) and the inverse document frequency (IDF) ((IDF(t, D))) scores. TF ((TF(t, d))) denotes the frequency of a term within a document, which can be assessed in several ways such as raw count, adjusted frequency relative to document length, logarithmically scaled frequency, or boolean frequency. Meanwhile, the Inverse Document Frequency (IDF) ($IDF(t, D)$) evaluates the scarcity of a term across the corpus. It is derived as the logarithm of the total number of documents divided by the number of documents that contain the term: The inverse document frequency (IDF) of a term (t) in a document set (D) is calculated as the logarithm of the ratio between the total number of documents (N) and the number of documents in which the term (t) appears [21]. We employed the TF-IDF method to extract a matrix of 10,000 relevant features from the financial news headlines. The extraction process involved utilizing various N-Gram ranges, spanning 1 to 3. This approach enhanced the analysis by capturing diverse contextual information contained within the text. This technology is extremely beneficial in converting text input into numerical representations, which improves the efficiency and accuracy of activities related to natural language processing and machine learning.

3.6 Classifiers

Linear SVC
The Linear SVC algorithm, which utilizes the 'crammer_singer' formulation, plays a crucial role in our financial news sentiment analysis [22, 23]. It aims to maximize a linear objective function while adhering to linear equality requirements. In a multi-class situation, the objective is to solve a problem by using a training set (X, y) consisting of input features X and target labels y.

$$\min_{w,b,\xi} \frac{1}{2}||w||^2 + C\sum_{i=1}^{n}\xi_i \tag{1}$$

subject to the constraints:

$$\forall_{i \neq y_i} : (w y_i \cdot x_i + b) - (w_i \cdot x_i + b) \geq \Delta(x_i, y_i, i) - \xi_i \tag{2}$$

where:

w is the weight vector.
b is the bias term.
ξ_i are slack variables to allow for misclassification.
C is the regularization parameter.
y_i and y_j are class labels.
Δ_{ij} is the margin, ensuring that the correct class score is higher than the scores of incorrect classes by a margin.

The objective of this formulation is to optimize the margins between different classes and minimize errors, hence enabling efficient sentiment categorization over a wide range of financial news categories. The application improves our ability to effectively identify and classify emotions within the complex realm of financial news headlines.

Logistic Regression

The LogisticRegression classifier, implemented with 'LogisticRegressionCV()', combines logistic regression with cross-validation to automatically discover the optimal regularization parameter (C) for improving sentiment analysis in financial news headlines [24]. Logistic regression utilizes predictor variables $((X_1, X_2, \ldots, X_p))$ and their corresponding coefficients $((\beta_0, \beta_1, \beta_2, \ldots, \beta_p))$ to estimate the probability ($P(Y = 1|X)$) for each sample.

'LogisticRegressionCV()' iteratively trains the model with different C values, controlling regularization, using the optimization problem:

$$\min_{\beta} \left(\frac{1}{n} \sum_{i=1}^{n} \log\left(1 + e^{-y_i(\beta_0 + \beta_1 X_{i1} + \beta_2 X_{i2} + \cdots + \beta_p X_{ip})}\right) + C \cdot ||\beta||^2 \right) \quad (3)$$

This process iteratively trains the model using different C values, which improves the model's adaptability for accurate sentiment classification in the complex field of financial news analysis [25].

Multinomial Naive Bayes

The Multinomial Naive Bayes classifier is a specialized algorithm used in our financial news sentiment analysis, tailored for text classification tasks. It operates on the assumption of independence between features (words) and calculates the probability of a specific sentiment category given the occurrence of certain words in the text [26].

Mathematically, it estimates the probability of a sentiment category c given a document d using Bayes' theorem:

$$P(c \mid d) = \frac{P(d \mid c) \cdot P(c)}{P(d)} \quad (4)$$

where:

$P(c \mid d)$ is the probability of category c given the document d.
$P(d \mid c)$ represents the likelihood of the document d occurring in category c.
$P(c)$ is the prior probability of category c.
$P(d)$ is the probability of document d.

The Multinomial Naive Bayes method employs word frequencies (or occurrences) to compute these probabilities. It counts the occurrences of each word in the documents belonging to specific sentiment categories and uses these frequencies to calculate the likelihood of a document's sentiment category based on the observed words [27].

This classifier's strength lies in its simplicity and effectiveness in handling text data by leveraging word frequency-based probabilities to categorize sentiments within financial news headlines accurately.

Random Forest

The Random Forest classifier, utilizing the 'RandomForestClassifier' algorithm with 100 estimators, plays a crucial role in our financial news sentiment analysis. It is a part of ensemble learning techniques, notably employing random forests, which combine predictions from several decision trees to enhance accuracy and dependability.

The approach operates by mathematically producing multiple decision trees throughout the training process. Each tree is built using a random subset of both the dataset and its attributes. The ultimate forecast is an amalgamation of forecasts derived from these varied decision trees. The process of integrating the predictions entails the computation of the average or aggregation of the outcomes derived from individual decision trees [28, 29]. The final prediction is calculated by taking the average of all individual predictions.

The formula for combining the predictions involves averaging or aggregating the outcomes from individual decision trees:

$$Prediction_{final} = \frac{1}{N}\sum_{i=1}^{N} Prediction_i \qquad (5)$$

where:

$Prediction_{final}$ represents the ultimate forecast.
$Prediction_i$ denotes the forecast generated by an individual decision tree.
N represents the aggregate number of decision trees in the forest.

The Random Forest Classifier with n_estimators = 100 explicitly creates a forest consisting of 100 decision trees. This method utilized in our financial news sentiment analysis leverages the combined predictive power of several trees, guaranteeing a strong comprehension of complex relationships within the data. It enhances the accuracy of sentiment classification by properly managing the intricacy of financial news feelings.

Gradient Boosting

The Gradient Boosting Classifier utilized in our financial news sentiment analysis employs Gradient Boosting, a technique that iteratively creates decision trees to minimize forecast errors [30]. The ultimate forecast is computed by amalgamating the weighted prognostications from each individual tree, gradually refining the overall outcome:

$$Prediction_{final} = Prediction_1 + \alpha_2 \times Prediction_2 + \alpha_3 \times Prediction_3 + ... \qquad (6)$$

where:

$Prediction_{final}$ represents the final prediction.
$Prediction_i$ denotes the prediction from an individual decision tree.
α_i signifies the weight assigned to each tree's prediction.

This iterative technique progressively enhances the precision of sentiment classification by minimizing errors in predictions. The Gradient Boosting Classifier improves sentiment analysis for financial news headlines by iteratively improving predictions using consecutive trees. This approach leads to greater accuracy and robustness in classification.

4 Results and Discussion

In our pursuit of precision in financial news headlines sentiment analysis, we conducted a comprehensive examination of various machine learning classifiers. This investigation delved into the influence of different N-Gram complexities Uni-Gram, Bi-Gram, and Tri-Gram on the accuracy of these classifiers. The Table 2 presents a detailed breakdown of the accuracy achieved by each classifier within these N-Gram models. Our focus lies in discerning the impact of N-Gram variations on multi-class classification accuracy, shedding light on the optimal approaches for sentiment analysis in financial news.

Table 2. Accuracy of Different Classifiers Using Various N-Gram Models.

Classifier\N-Gram	Uni-Gram	Bi-Gram	Tri-Gram
Linear SVC	**0.942**	**0.543**	0.365
Logistic Regression	0.927	0.543	**0.393**
Multinomial Naïve Bayes	0.814	0.532	0.391
Random Forest	0.904	0.524	0.388
Gradient Boosting	0.703	0.423	0.366

The Linear SVC model stands out as the top performance among the classifiers evaluated. It consistently achieves high accuracy rates across Uni-Gram, Bi-Gram, and Tri-Gram models, with significant scores of 94.2%, 54.3%, and 36.5% correspondingly. Logistic Regression exhibited a high level of accuracy, reaching 92.7%, in the analysis using Uni-Gram. However, when the complexity of N-Grams increased, its performance showed a declining trend. The initial accuracy of Multinomial Naïve Bayes was 81.4% when using Uni-Gram analysis. However, its efficacy significantly decreased while using more complex N-Gram models. Although Random Forest demonstrated impressive accuracy (90.4% for Uni-Gram), its performance declined significantly when dealing with more complex N-Grams (52.4% for Bi-Gram, 38.8% for Tri-Gram). On the other hand, Gradient Boosting showed a high level of accuracy of 70.3% while using Uni-Grams, but its performance decreased when dealing with more complex N-Grams (42.3% for Bi-Gram and 36.6% for Tri-Gram).

Our research emphasizes the importance of selecting models tailored to the complexities of financial news sentiment analysis, focusing on N-Gram sizes for multi-class classification. We discovered that model performance significantly improves when adapted to the unique features of financial texts, which often convey key sentiments subtly, challenging standard models.

We faced challenges including the variability of sentiment expression and the diverse formats of news sources, necessitating advanced NLP techniques for accurate interpretation and robust preprocessing pipelines to ensure data uniformity. Additionally, N-Gram models, while capturing immediate context, struggled with broader sentiment analysis and longer-term dependencies, indicating a need for models like transformers that consider wider textual contexts.

Despite these obstacles, our findings lay a foundation for future improvements in financial sentiment analysis, suggesting that model optimization could significantly enhance the detection and interpretation of nuanced sentiment indicators. Addressing these challenges and employing innovative modeling techniques are essential for advancing sentiment analysis precision and its application in financial market assessments.

5 Conclusion and Perspectives

Our exploration of sentiment analysis in financial news headlines through N-Gram models and diverse machine learning classifiers unveils promising findings and potential research paths. The noticeable variance in classifier performances across different N-Gram complexities underlines the complex dynamics of sentiment analysis within the financial domain. This investigation not only lays a foundational framework for the critical role of bespoke model selection in multi-class classification but also sets the stage for future advancements in developing more accurate prediction models.

Looking ahead, the prospect of hybrid models that meld various NLP techniques with sophisticated classifiers holds immense promise for enhancing the predictive accuracy in discerning the nuanced shifts in financial sentiment. Such advancements could lead to the creation of more nuanced analytical tools capable of accurately dissecting financial sentiments, thereby offering invaluable insights for investment decision-making and market forecasting. Moreover, the practical implementation of our findings could revolutionize the approach to financial market analysis, providing a robust toolset for investors and analysts to gauge market sentiments more accurately and make informed decisions. However, realizing these applications necessitates addressing computational resource challenges, especially for training and evaluating machine learning classifiers on expansive datasets. Future research should therefore also focus on optimizing computational efficiency, perhaps through more efficient algorithms or leveraging cloud computing resources, to facilitate the widespread adoption of these methodologies in real-world financial analysis scenarios.

In conclusion, while our study marks a significant step towards understanding and applying sentiment analysis in finance, it also opens up avenues for further exploration into advanced models and computational strategies. This progression will not only refine the precision of sentiment analysis tools but also broaden their applicability in practical financial decision-making contexts, ultimately contributing to more informed and strategic financial forecasting and investment practices.

References

1. Lin, H.-Y.: Efficient classifiers for multi-class classification problems. Decis. Support. Syst. **53**(3), 473–481 (2012). https://doi.org/10.1016/j.dss.2012.02.014
2. Dima, A., Lukens, S., Hodkiewicz, M., Sexton, T., Brundage, M.P.: Adapting natural language processing for technical text. Appl. AI Lett. **2**(3) (2021). https://doi.org/10.1002/ail2.33
3. Ferrario, A., Naegelin, M.: The art of natural language processing: classical, modern and contemporary approaches to text document classification. SSRN Electron. J. (2020). https://doi.org/10.2139/ssrn.3547887

4. Meyer, B., Bikdash, M., Dai, X.: Fine-grained financial news sentiment analysis. In: SoutheastCon 2017, Concord, NC, USA, pp. 1–8 (2017). https://doi.org/10.1109/SECON.2017.7925378
5. Mishev, K., Gjorgjevikj, A., Vodenska, I., Chitkushev, L.T., Trajanov, D.: Evaluation of sentiment analysis in finance: from lexicons to transformers. IEEE Access **8**, 131662–131682 (2020). https://doi.org/10.1109/access.2020.3009626
6. Kouissi, M., En-Naimi, E.M., Zouhair, A.: Tweets similarity classification based on machine learning algorithms, TF-IDF and the dynamic case based reasoning. In: Proceedings of the 6th International Conference on Networking, Intelligent Systems & Security, May 2023. https://doi.org/10.1145/3607720.3607796
7. Man, X., Luo, T., Lin, J.: Financial sentiment analysis (FSA): a survey. In: 2019 IEEE International Conference on Industrial Cyber Physical Systems (ICPS), Taipei, Taiwan, pp. 617–622 (2019). https://doi.org/10.1109/ICPHYS.2019.8780312
8. Agarwal, B.: Financial sentiment analysis model utilizing knowledge-base and domain-specific representation. Multimedia Tools Appl. **82**(6), 8899–8920 (2022). https://doi.org/10.1007/s11042-022-12181-y
9. Vicari, M., Gaspari, M.: Analysis of news sentiments using natural language processing and deep learning. AI & Soc. **36**(3), 931–937 (2020). https://doi.org/10.1007/s00146-020-01111-x
10. Financial News Headlines Data. Kaggle, 19 July 2020. www.kaggle.com/datasets/notlucasp/financial-news-headlines?select=cnbc_headlines.csv
11. Vijayarani, S., Ilamathi, Ms.J., Nithya, Ms.: Preprocessing techniques for text mining-an overview. Int. J. Comput. Sci. Commun. Netw. **5**(1), 7–16 (2015)
12. Kadhim, A.I.: An evaluation of preprocessing techniques for text classification. Int. J. Comput. Sci. Inf. Secur. (IJCSIS) **16**(6), 22–32 (2018)
13. Srividhya, V., Anitha, R.: Evaluating preprocessing techniques in text categorization. Int. J. Comput. Sci. Appl. **47**(11), 49–51 (2010)
14. Devitt, A., Ahmad, K.: Sentiment polarity identification in financial news: a cohesion-based approach. In: Proceedings of the 45th Annual Meeting of the Association of Computational Linguistics, pp. 984–991, June 2007
15. Jiao, J., Zhou, Y.: Sentiment polarity analysis based multi-dictionary. Phys. Procedia **22**, 590–596 (2011). https://doi.org/10.1016/j.phpro.2011.11.091
16. Sohangir, S., Petty, N., Wang, D.: Financial sentiment lexicon analysis. In: 2018 IEEE 12th International Conference on Semantic Computing (ICSC), pp. 286–289. IEEE January 2018
17. Bonta, V., Kumaresh, N., Janardhan, N.: A comprehensive study on lexicon based approaches for sentiment analysis. Asian J. Comput. Sci. Technol. **8**(S2), 1–6 (2019). https://doi.org/10.51983/ajcst-2019.8.s2.2037
18. Dey, A., Jenamani, M., Thakkar, J.J.: Senti-N-Gram: an n-gram lexicon for sentiment analysis. Expert Syst. Appl. **103**, 92–105 (2018). https://doi.org/10.1016/j.eswa.2018.03.004
19. Bespalov, D., Bai, B., Qi, Y., Shokoufandeh, A.: Sentiment classification based on supervised latent n-gram analysis. In: Proceedings of the 20th ACM International Conference on Information and Knowledge Management, pp. 375–382, October 2011
20. Qaiser, S., Ali, R.: Text mining: use of TF-IDF to examine the relevance of words to documents. Int. J. Comput. Appl. **181**(1), 25–29 (2018). https://doi.org/10.5120/ijca2018917395
21. Albitar, S., Fournier, S., Espinasse, B.: An effective TF/IDF-based text-to-text semantic similarity measure for text classification. In: Benatallah, B., Bestavros, A., Manolopoulos, Y., Vakali, A., Zhang, Y. (eds.) WISE 2014. LNCS, Part I, vol. 8786, pp. 105–114. Springer, Cham (2014). https://doi.org/10.1007/978-3-319-11749-2_8
22. Sikarwar, S.S., Tiwari, N.: Analysis the sentiments of amazon reviews dataset by using linear SVC and voting classifier. Int. J. Sci. Technol. Res. **9**(6), 461–465 (2020)

23. Sulaiman, S., Wahid, R.A., Ariffin, A.H., Zulkifli, C.Z.: Question classification based on cognitive levels using linear SVC. Test Eng. Manag. **83**, 6463–6470 (2020)
24. Jaya Hidayat, T.H., et al.: Sentiment analysis of twitter data related to Rinca Island development using Doc2Vec and SVM and logistic regression as classifier. Procedia Comput. Sci. **197**, 660–667 (2022). https://doi.org/10.1016/j.procs.2021.12.187
25. Majumder, S., Aich, A., Das, S.: Sentiment analysis of people during lockdown period of COVID-19 using SVM and logistic regression analysis. SSRN Electron. J. (2021). https://doi.org/10.2139/ssrn.3801039
26. Singh, G., Kumar, B., Gaur, L., Tyagi, A.: Comparison between multinomial and Bernoulli Naïve Bayes for text classification. In: 2019 International Conference on Automation, Computational and Technology Management (ICACTM), London, UK, pp. 593–596 (2019). https://doi.org/10.1109/ICACTM.2019.8776800
27. Abbas, M., Memon, K.A., Jamali, A.A., Memon, S., Ahmed, A.: Multinomial Naive Bayes classification model for sentiment analysis. IJCSNS Int. J. Comput. Sci. Netw. Secur. **19**(3), 62 (2019)
28. Karthika, P., Murugeswari, R., Manoranjithem, R.: Sentiment analysis of social media network using random forest algorithm. In: 2019 IEEE International Conference on Intelligent Techniques in Control, Optimization and Signal Processing (INCOS), Tamilnadu, India, pp. 1–5 (2019). https://doi.org/10.1109/INCOS45849.2019.8951367
29. Guia, M., Silva, R.R., Bernardino, J.: Comparison of Naïve Bayes, support vector machine, decision trees and random forest on sentiment analysis. KDIR **1**, 525–531 (2019)
30. Neelakandan, S., Paulraj, D.: A gradient boosted decision tree-based sentiment classification of Twitter data. Int. J. Wavelets Multiresolut. Inf. Process. **18**(04), 2050027 (2020). https://doi.org/10.1142/s0219691320500277

Comparative Study of Feature Selection Algorithms for Cardiovascular Disease Prediction with Artificial Neural Networks

Mohammed Marouane Saim(✉) and Hassan Ammor

ERSC Research Center, Mohammadia School of Engineers, Mohammed V University of Rabat, Rabat, Morocco
mohammedmarouanesaim@research.emi.ac.ma, ammor@emi.ac.ma

Abstract. Predicting cardiovascular diseases through advanced machine learning models is contingent upon the judicious selection of relevant features. This study engages in a comprehensive comparative analysis of eight feature selection algorithms, aiming to discern the optimal methodology for enhancing the predictive performance of an Artificial Neural Network (ANN) model. Leveraging the extensive dataset from the Framingham Heart Study, our investigation navigates through the intricacies of feature selection, evaluating algorithms ranging from correlation-based techniques to ensemble learning approaches. Each algorithm undergoes meticulous empirical evaluation, shedding light on its efficacy in distilling the dataset's complex features into a concise yet informative set. The foundational section of our study provides a detailed exploration of the Framingham dataset, outlining key attributes and their interpretations. Emphasizing the critical role of feature selection in predictive modeling, we elucidate the significance of refining datasets for optimal model performance. Our findings hold implications not only for the specific domain of cardiovascular health but also contribute to the broader landscape of predictive modeling in healthcare. The discernment of an optimal feature selection algorithm, tailored for ANNs, paves the way for more accurate, efficient, and interpretable models. This research offers valuable insights for healthcare professionals and researchers, augmenting the toolkit for advancing predictive capabilities in the realm of cardiovascular disease management.

Keywords: Cardiovascular Diseases · ANN · Feature Selection · Framingham

1 Introduction

Cardiovascular diseases remain a global health challenge, demanding innovative for effective prediction and management. At the forefront of this quest lies the integration of advanced machine learning techniques, particularly Artificial Neural Networks (ANNs), to decipher complex patterns inherent in health datasets. In this scientific exploration, we embark on a comparative journey, evaluating the efficacy of eight distinct feature selection algorithms. The primary objective is to discern the algorithm that optimally identifies crucial predictors for cardiovascular diseases, ultimately enhancing the predictive performance of an ANN model.

The foundation of our study rests upon the monumental Framingham Heart Study, a long-term investigation into the causes of heart disease within the community of Framingham, Massachusetts. The study's rich dataset encapsulates a myriad of attributes, providing a fertile ground for predictive modeling. Within this wealth of information lie potential predictors of cardiovascular diseases, waiting to be unearthed through meticulous feature selection.

Our endeavor unfolds across four interconnected sections, each unraveling a unique aspect of our comprehensive investigation. We commence by delving into the dataset from the Framingham study, outlining the key attributes and their interpretations. Subsequently, we illuminate the significance of feature selection in refining datasets for predictive modeling, setting the stage for the comparative analysis of eight feature selection algorithms. The chosen algorithms span diverse methodologies, ranging from correlation-based techniques to advanced ensemble learning approaches.

As we navigate through the empirical evaluation of these algorithms, our focus remains steadfast on their impact on the predictive prowess of an ANN model. Each algorithm undergoes rigorous scrutiny, shedding light on its ability to distill the dataset's intricate features into a compact yet informative set. The ultimate goal is to identify the algorithm that not only showcases optimal performance in isolating influential predictors but also harmonizes seamlessly with the complex architecture of an ANN model tailored for cardiovascular disease prediction.

This scientific pursuit not only contributes to the ever-evolving field of cardiovascular health prediction but also extends its implications to the broader landscape of predictive modeling in healthcare. By pinpointing the most effective feature selection algorithm for ANNs in this specific domain, we aim to pave the way for more accurate, efficient, and interpretable models, offering valuable insights for healthcare professionals and researchers alike.

2 Understanding the Dataset

The Framingham Heart Study dataset, extracted from the publicly accessible segment of the Framingham Heart Institute dataset, represents a cornerstone in unravelling the complexities underlying cardiovascular health within the Framingham, Massachusetts community. With a robust compilation of 4240 meticulously recorded instances, this dataset stems from a pioneering, longitudinal investigation spanning decades. This extensive research endeavour stands as a testament to its pivotal position within the domain of public health disease management, tirelessly probing the intricate interplay of factors contributing to cardiovascular ailments [1].

Comprising 14 discerning attributes meticulously curated for their profound implications on coronary heart disease, each feature encapsulates pivotal insights crucial for understanding the multifaceted nature of cardiovascular health [2]. These salient attributes encompass essential health metrics:

- Gender: Categorized dichotomously into Female (0) and Male (1), elucidating gender-based disparities in disease prevalence and risk factors.
- Age: Serving as a fundamental determinant, capturing the individual's age at the time of examination and acknowledging age-related implications in cardiovascular health.

- Current Smoker: A binary indicator (0 for non-smokers, 1 for smokers) reflecting the impact of smoking habits on cardiovascular health.
- Diabetes: A binary classification (0 for absence, 1 for presence) highlighting the substantial influence of diabetes in cardiovascular complications.
- Total Cholesterol (totChol): Quantifying the total cholesterol levels in milligrams per deciliters (mg/dL), an influential biomarker in assessing cardiovascular risk.
- Systolic Blood Pressure (sysBP): Measuring blood pressure during cardiac contraction, a critical determinant in cardiovascular health.
- Diastolic Blood Pressure (diasBP): Capturing blood pressure during cardiac relaxation, complementing systolic measurements in assessing overall cardiovascular health.
- Cigarettes Per Day (cigsPerDay): Offering insights into smoking intensity, an influential lifestyle factor in cardiovascular disease progression.
- BP Medications (BPMeds): Identifying individuals under blood pressure medication, signifying a crucial intervention in managing cardiovascular health.
- Prevalent Stroke: Noting occurrences of prevalent strokes, an impactful event significantly associated with cardiovascular complications.
- Prevalent Hypertension (prevalentHyp): Distinguishing prevalent hypertension cases, a pivotal risk factor contributing to cardiovascular diseases.
- Body Mass Index (BMI): Derived from the ratio of weight (in kilograms) to height (in meters squared), elucidating the weight-to-height correlation in cardiovascular health.
- Heart Rate: Measured in Beats Per Minute (Ventricular), reflecting the heart's functioning and its relevance to cardiovascular health assessment.
- Glucose: Quantifying glucose levels in milligrams per deciliter (mg/dL), a vital indicator of metabolic health and its association with cardiovascular risk factors.

Furthermore, a meticulous pre-processing step, leveraging information ranking theory, was methodically employed. This strategic approach facilitated the identification and retention of pivotal dataset variables, rigorously aligning the dataset with the study's central focus on discerning risk factors associated with coronary heart disease [3].

3 Importance of Feature Selection

Feature selection stands as a pivotal facet in the realm of machine learning, wielding a profound impact on model accuracy, interpretability, and generalization. This selective process of identifying and utilizing the most pertinent attributes from a dataset holds immense significance, particularly in the domain of predictive modeling for complex health outcomes such as cardiovascular diseases [4].

Significance in Machine Learning
Feature selection serves as a compass in navigating the intricate landscape of datasets, aiming to extract the most informative attributes while discarding redundant or irrelevant ones. By streamlining the feature space, models can focus on the most influential factors, enhancing their ability to discern patterns, make accurate predictions, and yield more interpretable outcomes. This process not only aids in mitigating overfitting—the model

learning noise in the data—but also fosters model transparency by spotlighting the most influential factors contributing to the predicted outcomes [5].

Improving Model Performance

- The inclusion of all available features in a dataset might seem intuitive but often leads to suboptimal model performance. Utilizing every attribute indiscriminately can introduce noise, increase computational complexity, and potentially hinder model generalization to unseen data. Through strategic feature selection, models can be fine-tuned to emphasize the most relevant aspects, enhancing their predictive power and robustness.

Challenges and the Need for Efficient Algorithms

- The abundance of features in real-world datasets poses a formidable challenge in model building. With a myriad of attributes at disposal, the computational burden escalates, and the risk of model overfitting amplifies. Moreover, irrelevant or redundant features might dilute the predictive capacity of the model, masking the true relationships between variables [6].

To navigate these challenges, efficient feature selection algorithms become indispensable. These algorithms traverse the feature space, evaluating attributes based on their predictive power or contribution to the model's performance. By strategically identifying and retaining the most informative features while discarding noise, these algorithms enhance model efficiency, interpretability, and generalization to unseen data.

4 Feature Selection Algorithms

In this study, eight distinct feature selection algorithms were rigorously evaluated to discern their efficacy in identifying crucial attributes for predicting coronary heart disease. Each algorithm was employed to extract informative features from the Framingham Heart Study dataset, facilitating a comparative analysis of their performance in enhancing model accuracy and interpretability.

Pearson Correlation Coefficient (PCC): This algorithm quantifies the linear relationship between attributes and the target variable. By calculating the correlation coefficients, it identifies attributes that showcase strong linear associations with the target, aiding in selecting features pivotal for predictive modeling [7].

Recursive Feature Elimination (RFE): RFE operates by iteratively removing less influential features based on model coefficients. This backward selection approach starts with the full feature set and sequentially prunes less significant attributes, refining the model's focus on the most informative variables [8].

Information Gain (IG): Assessing the information gain of each attribute concerning the target variable, Information Gain quantifies the reduction in uncertainty for predicting the target by including a particular attribute. Attributes showcasing higher information gain are considered more crucial for predictive modeling [9].

Chi-square Test: The Chi-square test evaluates the independence between attributes and the target variable in categorical datasets. This statistical method helps discern attributes significantly related to the target variable, aiding in feature selection for categorical data [10].

LASSO (L1 Regularization): LASSO, employing L1 regularization, encourages sparsity in the coefficient matrix. By penalizing less important attributes, it effectively selects relevant features, promoting a parsimonious yet informative model [11].

Random Forest Feature Importance (RF): Leveraging ensemble learning, Random Forest Feature Importance ranks feature importance based on their contribution to the model's predictive performance. This method identifies attributes crucial for accurate predictions within a decision tree-based ensemble framework [7].

Principal Component Analysis (PCA): PCA facilitates dimensionality reduction by transforming original attributes into a lower-dimensional space. By retaining maximal variance, PCA extracts orthogonal features, providing a condensed yet informative representation of the dataset [12, 13].

Recursive Feature Elimination with Cross Validation (RFECV): is a feature selection algorithm that uses recursive feature elimination and cross-validation to identify the most relevant features in a dataset. It operates by fitting the model, removing the least important features, and evaluating its performance through cross-validation. The algorithm assigns weights to features based on their predictive power, prunes the least significant ones, and repeats until the optimal subset is determined. Cross-validation ensures robustness by assessing performance across multiple data sets, mitigating overfitting. RFECV is a valuable tool for optimizing model performance and enhancing interpretability [14].

5 Methodology

The experimental design and evaluation framework play a pivotal role in assessing and comparing the efficacy of feature selection algorithms in predicting coronary heart disease using the Framingham Heart Study dataset. The methodology encompassed detailed procedures for dataset preparation, feature selection, model building, and performance evaluation.

Experimental Setup and Evaluation Metrics

- The study employed a systematic approach to evaluate the performance of feature selection algorithms. Key evaluation metrics encompassed standard measures such as accuracy, precision, recall, and F1-score. These metrics facilitated a comprehensive assessment of model performance concerning true positive, true negative, false positive, and false negative predictions.

Dataset Division and Pre-processing

- The Framingham Heart Study dataset, comprising 4240 records and 14 distinct attributes impacting coronary heart disease, underwent meticulous pre-processing.

As part of this process, an initial step involved the removal of redundant or irrelevant attributes to refine the dataset for subsequent analysis. Additionally, a feature selection approach, leveraging information ranking theory, was employed to identify and retain essential variables pertinent to the study's objectives.

The dataset was partitioned into distinct subsets for training and testing purposes. The division typically involved an 80/20, ensuring a balanced distribution of instances across both sets. The training set was utilized to train and validate the models, while the testing set served as an independent dataset to evaluate model generalization and performance on unseen data.

Model Building and Feature Selection
In the pursuit of crafting an effective predictive model for coronary heart disease, feature selection played a pivotal role in distilling the most influential attributes. Employing a suite of eight diverse feature selection algorithms, including Pearson Correlation Coefficient, Recursive Feature Elimination, and others, a systematic exploration unfolded on the training set. This meticulous process identified and retained attributes deemed most relevant for predicting coronary heart disease.

It's important to note that, in tandem with feature selection, a simplified Artificial Neural Network (ANN) model, comprising merely two nodes, was employed. This minimalist ANN configuration aimed to assess the impact of the selected features on a basic neural network architecture. This conjoint approach sought not only to discern the most effective feature selection algorithms but also to understand how these features contribute to the predictive capacity of a simple ANN model. The models, encompassing logistic regression, decision trees, or ensemble methods, were subsequently constructed using the gleaned insights from feature selection. Rigorous fine-tuning and optimization ensued within the confines of the training dataset to ensure optimal model performance. This dual strategy aimed to harness the strengths of both feature selection algorithms and a simplified ANN model for a nuanced exploration of predictive modeling in the context of coronary heart disease [15].

Performance Evaluation
The models generated through different feature selection algorithms underwent rigorous evaluation on the dedicated testing set. Accuracy is calculated to assess the models' predictive capabilities [10]. This comprehensive evaluation facilitated the comparison of model performance across different feature selection techniques, aiding in identifying the most effective approach for predicting coronary heart disease.

6 Results and Analysis

The results section consists of two parts, each providing insights into the effectiveness of feature selection algorithms and their impact on predictive modeling. The first segment details the outcomes of eight feature selection algorithms, revealing the nuances of feature importance rankings across different methodologies. The second segment examines the performance of the Artificial Neural Network (ANN) model, evaluating the efficacy of individual feature selection techniques and their impact on predictive accuracy.

6.1 Feature Selection Outcomes

The application of diverse feature selection algorithms to the Framingham Heart Study dataset has offered valuable insights into potential predictors of coronary heart disease. Each algorithm identified distinct sets of attributes, shedding light on the variables deemed most influential in the context of this study (Table 1).

Table 1. Selected Features and Ranking by Feature Selection Algorithms

	PCC	REF	IG	Chi-Square	LASSO	RF	PCA	RFECV
Gender	7	–	–	–	6	–	–	1
age	1	1	1	1	1	3	3	2
currentSmoker	–	–	–	–	7	–	–	–
cigsPerDay	–	2	–	2	3	8	–	3
BPMeds	–	–	–	–	–	–	–	–
prevalentStroke	–	–	4	–	–	–	–	–
prevalentHyp	3	–	3	3	–	–	7	–
diabetes	6	–	–	4	–	–	–	–
TotChol	8	3		5	5	4	1	4
SysBP	2	4	2	6	2	1	2	5
DiaBP	4	5	3	7	–	5	3	6
BMI	–	6	6	–	–	2	6	7
HeartRate	–	7	5	–	–	7	5	–
Glucose	5	8	–	8	4	6	4	–

Observations:

– Consistency Across Algorithms:

Features such as 'age' and 'sysBP' consistently appear among the top-ranking features across various algorithms, suggesting their robust influence on predicting coronary heart disease.

– Divergence in Prioritization:

While certain features exhibit consistency, the algorithms diverge in their prioritization of other variables. For instance, 'glucose' is highly ranked by Pearson Correlation Coefficient and Information Gain but ranks lower in Recursive Feature Elimination.

– Algorithm-specific Emphasis:

Each algorithm emphasizes specific aspects of the dataset. Random Forest Feature Importance highlights 'sysBP' and 'BMI,' indicating their pronounced impact. On the other hand, PCA focuses on 'totChol' and 'prevalentHyp' as key predictors.

Implications:

– Feature Redundancy:

The consistency in certain features across algorithms suggests their importance in predicting coronary heart disease. Redundant features may be identified for potential removal, streamlining the model without sacrificing predictive accuracy.

– Algorithm Selection Considerations:

The choice of a feature selection algorithm should align with the goals of the predictive modeling task. Random Forest Feature Importance, for example, provides insight into feature importance in an ensemble context, while PCA offers dimensionality reduction benefits.

– Interpretability vs. Complexity:

The divergence in feature rankings underscores the trade-off between model interpretability and complexity. Researchers must carefully weigh the interpretative value of selected features against the computational demands of the chosen algorithm.

6.2 Predictive Modeling with Selected Features

The predictive models built using features selected by each algorithm were evaluated based on accuracy (Fig. 1).

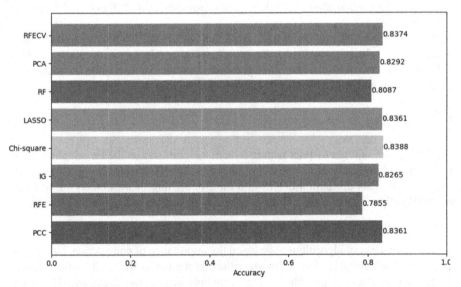

Fig. 1. Accuracy of ANN model with different feature selection algorithms

- Pearson Correlation Coefficient (PCC): excels at capturing linear correlations, emphasizing the significance of age, blood pressure, diabetes, and cholesterol levels in predicting coronary heart disease. The algorithm's robust performance underscores the value of straightforward correlations in feature selection.
- Recursive Feature Elimination (RFE): emphasizes the importance of age, smoking habits, cholesterol levels, and blood pressure in predicting coronary heart disease. Its robustness in aligning with linear correlations provides a comprehensive view of influential factors.
- Information Gain (IG): prioritizes attributes that reduce uncertainty, highlighting the crucial contributions of age, blood pressure, prevalent hypertension, and heart rate to accurate predictions.
- Chi-square Test: effective for categorical data, underscores the significance of age, smoking, prevalent hypertension, diabetes, and cholesterol levels as key contributors to predicting coronary heart disease.
- LASSO (L1 Regularization): emphasis on sparsity yields concise feature sets, emphasizing the importance of age, blood pressure, smoking intensity, and glucose levels in predictive modeling.
- Random Forest Feature Importance (RF): leveraging ensemble learning, identifies key features such as blood pressure, body mass index, age, and glucose levels, demonstrating the significance of these factors in predicting coronary heart disease within the ensemble framework.
- Principal Component Analysis (PCA): PCA's ability to reduce dimensionality highlights the importance of cholesterol levels, blood pressure, age, and glucose levels in a lower dimensional space, providing insights into key contributors to coronary heart disease.
- Recursive Feature Elimination with Cross Validation (RFECV): dynamically adjusting feature subsets, emphasizes the importance of gender, age, education, smoking habits, cholesterol levels, and blood pressure in predictive modeling, showcasing its adaptability and effectiveness in capturing diverse influential factors.

7 Discussion

The exploration of eight distinct feature selection algorithms on the Framingham Heart Study dataset has unveiled a rich tapestry of insights into potential predictors for coronary heart disease. This discussion delves into the overarching themes emerging from the results, highlighting the implications for predictive modeling in healthcare applications.

– Algorithmic Diversity and Feature Importance:

The diverse set of algorithms showcased the importance of considering various perspectives in feature selection. Each algorithm, grounded in different methodologies, presented a unique lens through which potential predictors were identified. This diversity contributes to a more holistic understanding of the complex interplay of factors influencing coronary heart disease.

– Robust Predictors Across Multiple Algorithms:

Notably, certain features, such as age, blood pressure (systolic and diastolic), and cholesterol levels, consistently emerged as robust predictors across multiple algorithms. The consistent selection of these attributes underscores their fundamental role in predictive modeling for coronary heart disease and aligns with existing medical knowledge.

– Sparse Models vs. Ensemble Learning:

LASSO (L1 Regularization) emphasized sparsity, producing concise feature sets, while Random Forest (RF) leveraged ensemble learning to identify key features. This juxtaposition highlights the trade-off between simplicity and complexity in model interpretability, with LASSO favoring simplicity and RF embracing the collective power of multiple features.

– Dimensionality Reduction and Interpretability:

Principal Component Analysis (PCA), with its ability to reduce dimensionality, provided insights into features that contribute significantly to coronary heart disease in a lower-dimensional space. This reduction, while aiding interpretability, also raises questions about the potential loss of nuance inherent in high-dimensional data.

– Adaptability and Dynamic Feature Subsets:

Recursive Feature Elimination with Cross-Validation (RFECV) showcased adaptability by dynamically adjusting feature subsets. The algorithm emphasized the importance of gender, age, education, smoking habits, cholesterol levels, and blood pressure, highlighting its effectiveness in capturing diverse influential factors while adjusting to the intricacies of the dataset.

– BPMeds and Algorithmic Oversight:

A notable observation is the absence of Blood Pressure Medications (BPMeds) in the selected features across all algorithms. This may suggest that, within the context of the Framingham Heart Study dataset and the specific algorithms applied, BPMeds might not be a primary predictor for coronary heart disease. However, caution is warranted, and further investigation is needed to validate this observation.

The results and discussions presented herein contribute to the ongoing discourse on feature selection methodologies in healthcare applications. The insights gained from these algorithms provide a foundation for future research, emphasizing the need for a nuanced understanding of both the chosen algorithmic approach and the specific characteristics of the dataset under consideration. The continual refinement and validation of feature selection strategies are crucial for advancing predictive modeling in the realm of cardiovascular health.

8 Conclusion

This study embarked on a comprehensive exploration of feature selection algorithms within the context of the Framingham Heart Study dataset, aiming to unravel the intricate web of predictors for coronary heart disease. The diverse set of eight algorithms, each wielding its unique approach, provided nuanced perspectives on feature importance and model interpretability.

The robust predictors consistently identified across multiple algorithms, including age, blood pressure, and cholesterol levels, underscore their fundamental role in predictive modeling for coronary heart disease. This alignment with established medical knowledge enhances the credibility of our findings and reinforces the importance of these factors in understanding cardiovascular health.

The dichotomy between sparse models, exemplified by LASSO (L1 Regularization), and the ensemble approach of Random Forest (RF) showcased the perpetual trade-off between simplicity and complexity in model design. While LASSO offered concise feature sets, RF harnessed the collective strength of multiple features. The interpretability and computational efficiency of these approaches necessitate careful consideration in the broader context of healthcare applications.

Principal Component Analysis (PCA) illuminated the dimensionality reduction capabilities, shedding light on features contributing significantly in a lower-dimensional space. This reduction aids interpretability but prompts reflections on potential information loss, urging a balance between simplicity and data richness.

Recursive Feature Elimination with Cross-Validation (RFECV) demonstrated adaptability, dynamically adjusting feature subsets. Its emphasis on diverse factors such as gender, education, and smoking habits highlighted the algorithm's efficacy in capturing the intricate relationships within the dataset.

Notably, the absence of Blood Pressure Medications (BPMeds) in the selected features prompts further inquiry into the role of medications in predictive modeling. This underscores the need for continual refinement and validation of feature selection strategies, acknowledging the evolving nature of healthcare data.

In conclusion, this study contributes to the evolving landscape of feature selection methodologies in predicting coronary heart disease. The findings offer valuable insights for researchers and practitioners, emphasizing the importance of algorithmic diversity, model interpretability, and the nuanced understanding of dataset characteristics. As we navigate the complexities of cardiovascular health prediction, this exploration lays the groundwork for future endeavors in refining predictive models and advancing public health interventions.

References

1. Andersson, C., Nayor, M., Tsao, C.W., Levy, D., Vasan, R.S.: Framingham heart study: JACC focus seminar, 1/8. J. Am. Coll. Cardiol. **77**(21), 2680–2692 (2021)
2. O'Donnell, C.J., Elosua, R.: Cardiovascular risk factors. Insights from framingham heart study. Revista Española de Cardiología (Engl. Edn.) **61**(3), 299–310 (2008)
3. Jalepalli, S.K., et al.: Development and validation of multicentre study on novel artificial intelligence-based cardiovascular risk score (AICVD). Fam. Med. Commun. Health **12**(Suppl. 1), e002340 (2024)
4. Büyükkeçeci, M., Okur, M.C.: A comprehensive review of feature selection and feature selection stability in machine learning. Gazi Univ. J. Sci., 1 (2024)
5. Khan, M.N.A., et al.: Prediction of hydrogen yield from supercritical gasification process of sewage sludge using machine learning and particle swarm hybrid strategy. Int. J. Hydrog. Energy **54**, 512–525 (2024)

6. DeGroat, W., et al.: Discovering biomarkers associated and predicting cardiovascular disease with high accuracy using a novel nexus of machine learning techniques for precision medicine. Sci. Rep. **14**(1), 1 (2024)
7. Liu, S., et al.: Continuous blood pressure monitoring using photoplethysmography and electrocardiogram signals by random forest feature selection and GWO-GBRT prediction model. Biomed. Signal Process. Control **88**, 105354 (2024)
8. Manikandan, G., et al.: Classification models combined with Boruta feature selection for heart disease prediction. Inform. Med. Unlocked, 101442 (2024)
9. Mishra, J.S., et al.: Evaluating the effectiveness of heart disease prediction. Int. J. Intell. Syst. Appl. Eng. **12**(5s), 163–173 (2024)
10. Mehari, T., et al.: ECG feature importance rankings: cardiologists vs. algorithms. IEEE J. Biomed. Health Inform. (2024)
11. Kalaivani, B., Ranichitra, A.: Optimizing cardiovascular disease prediction: harnessing random forest algorithm with advanced feature selection (2024)
12. Jolliffe, I.T.: Principal Component Analysis, 2nd edn. Springer, New York (2002)
13. Greenacre, M., Groenen, P.J.F., Hastie, T., Iodice D'Enza, A., Markos, A., Tuzhilina, E.: Principal component analysis. Nat. Rev. Methods Primers **2**, 100 (2022). https://doi.org/10.1038/s43586-022-00184-w
14. Zhou, S., et al.: Interpretable machine learning model for early prediction of 28-day mortality in ICU patients with sepsis-induced coagulopathy: development and validation. Eur. J. Med. Res. **29**(1), 1–14 (2024)
15. Parashar, G., Chaudhary, A., Pandey, D.: Machine learning for prediction of cardiovascular disease and respiratory disease: a review. SN Comput. Sci. **5**(1), 196 (2024)

TinyML and Anomaly Detection

Ensemble Learning for Malware Detection

Loubna Moujoud[(✉)], Meryeme Ayache, and Abdelhamid Belmekki

Department of Mathematics Computer Science and Networks, INPT Rabat, Rabat, Morocco
{moujoud.loubna,ayache,belmekki}@inpt.ac.ma

Abstract. One of the major threats to security on the Internet nowadays is Malware. In fact, malware is frequently the root cause of the majority of Internet issues, including phishing and APT (Advanced Persistent Threat) attacks [1]. New methods have been proposed for detecting and preventing the most recent generation of malware. Nevertheless, there hasn't been much research done on how effectively ensemble approaches perform. In this paper, we have designed two approaches based on ensemble learning to detect malware. The first approach uses a stacking method to create a set of machine learning classifiers, whereas the second approach combines several deep learning algorithms. For comparison, six individual machine learning algorithms and two deep learning algorithms were used. A dataset of 42,797 malware API call sequences and 1,079 goodware API call sequences is used for the evaluation of the suggested methods. The results of the experiments validate the proposed methods' performance.

Keywords: Malware · Malware Detection · Classification · Machine Learning · Deep Learning · Ensemble Learning

1 Introduction

Any malicious program designed by hackers to steal data and damage or destroy computers and systems is known as malware. Worms, which are self-replicating software programs, were the first example of malware. Before networks were widely used, computer worms propagated through contaminated storage media. One of the earliest examples of a malware attack in history is Melissa [2], which was developed in 1999. A word document had a macro embedded in it. When the user opens the file, the macro will launch and infect the first 50 contacts in the user's address book. Similarly, multiple virus attacks throughout history include, My Doom worm in 2004 [3], Stuxnet in 2010 [4], WannaCry in 2017 [5], Emotet in 2018 [6], and CovidLock in 2020 [7]. Malware has been categorized into many different types [8], some types are illustrated in Fig. 1.

Early in the cyber era, there were comparatively few malware threats, and pre-execution rules that were manually written were frequently sufficient for threat detection. With the explosive expansion of malware and the Internet,

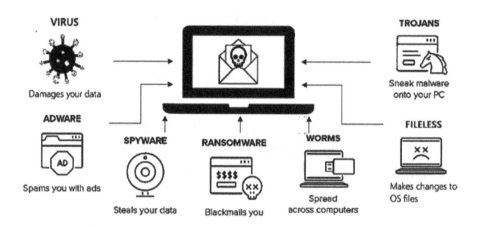

Fig. 1. Types of Malware

manually developed detection rules became impractical, and new techniques were needed.

Machine learning and deep learning have emerged as powerful tools in the realm of malware detection. These technologies enable the development of robust, adaptive, and proactive defense mechanisms against the ever-evolving landscape of malware threats. Machine learning algorithms can analyze vast datasets, identifying patterns and anomalies in file attributes, network behavior, and system processes. Deep learning, a subset of machine learning, employs complex neural networks to automatically extract intricate features and recognize subtle signatures of malware. Their ability to learn and evolve makes them adept at detecting both known and novel malware strains, offering a critical line of defense in safeguarding data and systems from malicious encryption and extortion attempts. As malware attacks become increasingly sophisticated, the integration of machine learning and deep learning techniques in cybersecurity plays a vital role in staying one step ahead of these cyber threats.

Ensemble learning has proven to be a powerful and effective approach in the realm of cybersecurity, particularly in the detection of malware. By combining multiple diverse classifiers, ensemble methods enhance overall detection accuracy and robustness. An ensemble architecture may involve a variety of machine learning and deep learning algorithms, such linear regression, random forests, support vector machines, and neural networks. Due to this diversity, the ensemble is better able to generalize and capture a wider variety of malware features, increasing its resistance against emerging threats.

In this paper, our primary goal is to provide a comparison between machine learning, deep learning and ensemble learning in malware detection. The main contributions are summarized as below:

1. Using Scikit-learn resources, which offers a large selection of machine learning techniques;
2. To detect and classify malware, two ensemble learning techniques based on machine learning and deep learning algorithms are presented;
3. On a dataset of 42,797 malware API call and 1,079 goodware API call, experiments are performed to evaluate the performance of suggested ensemble algorithms, and findings are compared with other individual classifiers.

The paper is organized as follows: The first section is dedicated to the introduction. Section 2 covers the background of ensemble learning and an overview of the existing works. Section 3 presents our proposed work. Section 4 is a discussion section that provides the results and the findings of our work, and followed by Sect. 5 for conclusion.

2 Background and Related Work

2.1 Background

Ensemble methods are meta-algorithms that combine multiple machine learning and deep learning techniques into a single predictive model, in order to reduce variance (bagging), bias (boosting), or enhance predictions (stacking) [9]. There are two types of ensemble learning: Sequential ensemble methods, in which the basis learners are generated sequentially, and parallel ensemble methods, in which the base learners are generated simultaneously. To create homogenous base learners, or learners of the same type, the majority of ensemble methods employ a single base learning algorithm. Additionally, some approaches employ heterogeneous learners, or learners of several types. To ensure that ensemble approaches outperform any of its individual members, the base learners must be as diverse and precise as possible.

1. **Bagging**: is an ensemble learning method that enhances a machine learning model's prediction performance by combining the advantages of aggregation and bootstrapping to produce a stable model. The primary goal of bagging is to lower a dataset's variance, which guarantees the model's stability and prevents it from being impacted by certain dataset samples.
2. **Boosting**: is an ensemble learning method that transforms weak learners into strong learners. Boosting works by fitting a series of weak learners (models, like small decision trees, that are just marginally better than random guessing) to weighted versions of the data, where samples that were incorrectly classified in previous rounds are assigned extra weight. The final forecast is then obtained by combining the predictions using either a weighted sum (regression) or a weighted majority vote (classification).

3. **Stacking**: is an ensemble learning method that employs a meta-classifier or meta-regressor, to merge numerous regression or classification models. Once the base level models have been trained using the whole training set, the meta-model is trained using the outputs of the base level model as features. Stacking ensembles are usually diverse because several methods of learning are often applied at the base level.

The distinction between ensemble learning and machine/deep learning lies in their scope and focus:

1. **Scope**
   - Machine learning and deep learning are more general domains that include a range of methods and algorithms that allow algorithms see patterns and decide what to predict or perform without explicit programming.
   - Ensemble learning is a technique that combines several models, machine learning or deep learning models, to improve overall performance.
2. **Focus**
   - Machine/deep learning's main goal is to give computers the experience with data they need to learn and become more proficient at a particular task.
   - Ensemble learning leverages the strengths of multiple models to enhance predictive accuracy, robustness, and generalization, and compensate for the weaknesses of individual models.

2.2 Related Work

Algorithms can detect an object's digital signature rapidly and effectively by scanning it. A known malware database is updated with an object's signature when an anti-malware solution provider determines it to be malicious. Hundreds of millions of signatures that reveal suspicious items may be included in these repositories. The main advantages of the signature-based malware detection technique are its speed, ease of use, and widespread availability. Its incapacity to identify unidentified attacks is one of its main drawbacks. To avoid being discovered, malicious actors can easily alter the attack sequences they use in malware and other forms of attacks. Yerima et al. [10] proposed a method for android malware detection, that combines the benefits of static analysis with the effectiveness and performance of ensemble machine learning. A well-known antivirus collection of good and malicious samples has been used to build the machine learning models. The suggested method could achieve 97.3–99% detection accuracy with very low false positive rates.

While behavior-based malware detection checks an object planned actions before it can actually execute that behavior. Many different behaviors could indicate possible danger. For example: Attempts to find a sandbox environment, turning off security measures, installing rootkits, and setting up AutoStart. Unknown malware can be detected more effectively with behavior-based detection, but because it demands running the malware in an isolated environment, it is frequently slow and expensive. Feng et al. [11] provided an framework for dynamic analysis, named EnDroid, based on a variety of dynamic behavior features. These features include common application-level malicious actions like stealing personal information, subscribing to premium services, and communicating with malicious services, as well as system-level behavior tracing. EnDroid uses an ensemble learning method to extract behavior data from runtime monitors in order to differentiate between malicious and good programs. The efficacy of EnDroid on two datasets has been demonstrated through experiments. Moreover, stacking works the best in terms of classification.

The technology of cloud computing is expanding daily, but it also brings with it a number of security challenges, particularly from malware and other dangers. Multiple researchers offer surveys that address particular cloud security issues and suggested improvements. Li et al. [12] suggested a technique that builds a detection model by combining the Bagging and Boosting algorithms to find illicit mining code on cloud platforms. The variance of model detection can be significantly decreased by randomly extracting samples and allowing models to vote collectively to make a decision. When compared to conventional classifiers, the suggested approach can achieve superior robustness and accuracy. According to the experimental findings, the AUC (Area Under the Curve) and F1-score values for the provided dataset can, respectively, reach 0.992 and 0.987.

Malware attacks can affect mobile systems just like they can any other information system. Due to the unique characteristics of mobile devices, malware detection is an essential feature that every device must have in order to safeguard sensitive information and prevent attacks. According to Skycure [13], Android smartphones are almost twice as likely as iOS devices to carry malware, and one-third of mobile devices have a medium to high risk of data being exposed. Christianah et al. [14] proposed an ensemble learning to create an optimum model for Android virus detection. Three different base models were created using Random Forest, Support Vector Machine, and k-Nearest Neighbors. The prediction outcomes of these models were merged using the Majority Vote combination function to create an ensemble model. To extract static features from an extensive selection of malware samples and goodware apps, reverse engineering was used. According to the results, Random Forest is a good base model since it can accurately categorize instances 98% of the time. Another approach [15] is suggested by the authors to identify Android malware, and they combined a variety of permissions and intentions with Ensemble techniques. Applying the suggested method to 1,745 real-world applications achieves 99.8% accuracy, which is the highest recorded accuracy so far.

Furthermore, IoT technologies will be an endless rise in malware attacks. The security community has directed its research efforts in the last few years toward the detection of malware on IoT devices. The Internet of Things (IoT) has been implemented by over 80% of companies, and in the last three years, 20% of them have identified an IoT-based attack [16]. Several studies were introduced to detect IoT malware. Vasan et al. [17] developed an efficient cross-architecture threat hunting model for IoT malware using advanced ensemble learning (MTHAEL). A stacked ensemble of heterogeneous feature selection algorithms is used by the model. The model is thoroughly tested on a big IoT cross-architecture dataset consisting of 21,137 samples. It outperforms previous related research with an accuracy of 99.98% for classifying ARM architecture samples.

Then, we have deep learning algorithms, such as convolutional neural networks (CNN), recurrent neural networks (RNN) and long-short term memory (LSTM), which have been applied to a several use cases in malware detection. Yan et al. [18] suggested MalNet, an advanced malware detection technique that automatically extracts features from unprocessed data. MalNet learns from grayscale images and opcode sequences using CNN and LSTM networks, respectively, and uses a stacking ensemble for malware classification. More than 40,000 samples were used for their studies, comprising 21,736 malicious files supplied by Microsoft and 20,650 benign files gathered from internet software vendors. MalNet obtained 99.88% validation accuracy for malware detection.

3 Proposed Work

3.1 Dataset

The dataset on which the model was built was hosted on the Kaggle portal [19]. It contains 42,797 malware API call sequences and 1,079 goodware API call sequences. Using Cuckoo Sandbox reports, the first 100 consecutive, non-repeated API calls linked to the parent process are obtained for every software.

- **Hash**: MD5 hash of the sample
- **Malware**: 0 for Goodware & 1 for Malware
- [t_0 ... t_99]: API calls

	hash	t_0	t_1	t_2	t_3	t_4	t_5	t_6	t_7	t_8	...	t_91	t_92	t_93	t_94	t_95	t_96	t_97	t_98	t_99	malware
0	071e8c3f8922e186e57548cd4c703a5d	112	274	158	215	274	158	215	298	76	...	71.0	297.0	135.0	171.0	215.0	35.0	208.0	56.0	71.0	1.0
1	33f8e6d08a6aae939f25a8e0d63dd523	82	208	187	208	172	117	172	117	172	...	81.0	240.0	117.0	71.0	297.0	135.0	171.0	215.0	35.0	1.0
2	b68abd064e975e1c6d5f25e748663076	16	110	240	117	240	117	240			...	65.0	112.0	123.0	65.0	112.0	123.0	65.0	113.0	112.0	1.0
3	72049be7bd30ea61297ea624ae198067	82	208	187	208	172	117	172	117	172	...	208.0	302.0	208.0	302.0	187.0	208.0	302.0	228.0	302.0	1.0
4	c9b3700a77facf29172f32df6bc77f48	82	240	117	240	117	240	117	240	117	...	209.0	260.0	40.0	209.0	260.0	141.0	260.0	141.0	260.0	1.0

Fig. 2. Dataset

A set of 100 API calls explaining each process are included in each row. Software activity, for example, could appear like this: create new folder, create new file, edit file, save file, etc.; these calls are recorded in order, and the final column classifies the activity as malware or goodware.

3.2 Feature Selection

The Extra Trees Classifier [20], which helps choose features based on their relevance scores and lowers sensitivity to noise and irrelevant features, extracted the chosen features. It may accelerate the classification of samples as malware or goodware by efficiently handling datasets with a high number of attributes and noisy data. In our case, the Extra Trees Classifier determined that 27 features were required.

3.3 Data Classification

The performance of ensemble learning can be increased by including a diversity of classifier types, each of which employs a particular approach for classifying the data. The learning model is able to attain accuracy that would be impossible for any one of the individual classifiers. As base classifiers, we have employed five machine learning classifiers and two deep learning classifiers, namely, Random Forest, Support Vector Machine, Decision Tree, K-Nearest Neighbors, Logistic Regression, Convolutional Neural Network & Long Short-Term Memory. These classifiers are assessed using a range of evaluation factors. Below is a description of each of these base classifiers:

Machine Learning Algorithms

- **Random Forest (RF)** Random forest is a popular machine learning algorithm, which combines the results of multiple decision trees to get a single result. Its popularity has been aided by its adaptability, simplicity of usage, and capacity to handle both regression and classification issues.
- **Support Vector Machine (SVM)** Support Vector Machine is a simple method used for classification and regression, and can achieve high classification accuracy. But it is only appropriate for small samples, a large sample set will result in a high false positive rate and significant delay in processing.
- **Decision Tree (DT)** Decision Tree is a non-parametric technique that determines the most effective split to measure the homogeneity of the variables within each subgroup of a dataset.
- **K-Nearest Neighbors (KNN)** The KNN algorithm is a non-parametric supervised learning classifier that uses proximity to classify or predict a group of single data point. It is typically used as a classification technique, depending on the theory that similar points can be found adjacent to one another.
- **Logistic Regression (LR)** Logistic Regression is usually employed for binary category issues. It calculates the possibility that an event will occur. Considering a probability as a result, the dependent variable's range is 0 to 1.

Deep Learning Algorithms

- **Convolutional Neural Network (CNN)** A CNN is a type of feed-forward neural network in which the positioning of its neurons is modeled like the structure of the animal visual cortex. It learns more quickly and makes less errors.
- **Long Short-Term Memory (LSTM)** LSTM networks are a type of recurrent neural network (RNN), made to identify long-term dependencies in sequential data. Memory cells and gates in LSTM networks enable them to selectively recall or erase information over time.

3.4 Proposed Methods

First method proposed in this paper is an ensemble of machine learning algorithms. We applied the stacking method to combine these classifiers: RF, SVM, DT & KNN. We employed LR as a meta-classifier. The data instances are categorized into a set of classes using the LR method. The logistic sigmoid function is used to convert the output, and the result is a probability value that can be easily applied to other classes.

Second method is mainly based on a deep learning ensemble that combines CNN and LSTM networks to detect malware, by stacking several layers in a hierarchical way. CNN has shown interest in the detection of malware due to their capacity to comprehend the underlying patterns and traits that are predictive of suspicious activities. In our model, we employed the output of the CNN as the input to the LSTM. As a result, the CNN's learnt features from the input data can be learned by the LSTM. Both methods are presented in the Fig. 3.

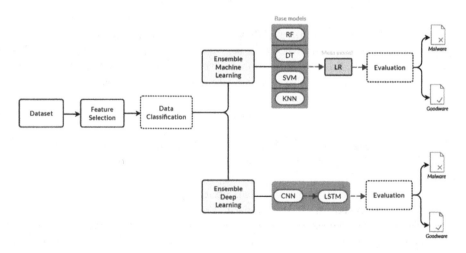

Fig. 3. Proposed system for malware detection

In both methods, binary classification was performed on each training and testing set for both malware and goodware samples. Performance on all features was conducted as well as additional training and testing from selected features. Data split is created. As a results, training and testing datasets are created. Both of them include a very similar proportion of malware and goodware observations (training: 19222 goodware & malware while testing: 19223). The following modules, scores, and classifiers were imported from a scikit python learning module for testing and training. In Sect. 4, we will see how well these methods can perform.

4 Results and Discussion

This section outlines the aspects of the evaluation, and the results of the suggested methods. The final results are compared with the base classifiers. We used the model's accuracy and F1-score as performance indicators.

4.1 Performance Indicators

We chose a 50% to 50% training to testing ratio. We employed two evaluation metrics: F1-score & accuracy. The proportion of malware and goodware samples are appropriately categorized as True Negatives (TNs) and True Positives (TPs). Similarly, a considerable amount of goodware and malware samples are falsely categorized as False Negatives (FNs) and False Positives (FPs). The following metrics are performance comparison metrics of each of the algorithms used:

$$Accuracy = \frac{TP + TN}{TP + TN + FP + FN} \tag{1}$$

$$F1 = \frac{2 * \frac{TP}{TP+FP} * \frac{TP}{TP+FN}}{\frac{TP}{TP+FP} + \frac{TP}{TP+FN}} \tag{2}$$

4.2 Experiment Results

The comparison of the nine malware detection performance measures is shown in Tables 1 and 2. The LSTM model has the lowest performance with 98,78% & 97,59%, F1-score and accuracy respectively. However, the ML ensemble learning model (RF+SVM+DT+KNN+LR) performs best in terms of F1-score & accuracy, with 99,95% & 99,91% respectively. While the DL ensemble learning model (CNN+LSTM) comes in second place with 99,37% & 98,75% in F1-score & accuracy.

Figures 4 and 5 help us visualize the comparison of F1 score and accuracy of the ensemble models and individual models. According to findings, ensemble learning outperforms traditional machine learning and deep learning algorithms, in terms of accuracy and F1 score, and effective to detect malware.

The following conclusions can be drawn from the analysis based on the evidence that was previously presented:

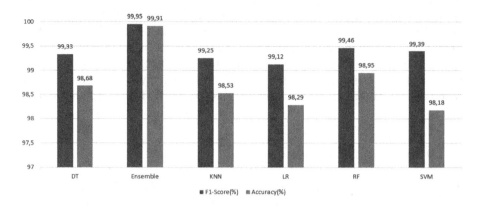

Fig. 4. Comparison of ML classifiers on the basis of F1 score & Accuracy

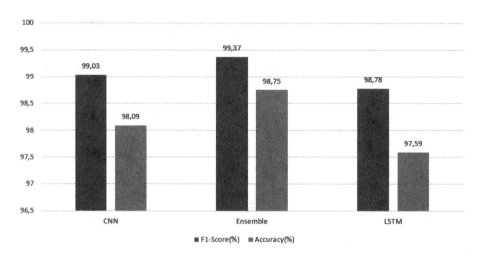

Fig. 5. Comparison of DL classifiers on the basis of F1 score & Accuracy

Table 1. Comparison of performance measures using machine learning

Model	F1-Score (%)	Accuracy (%)
DT	99,33	98,68
LR	99,12	98,29
RF	99,46	98,95
KNN	99,25	98,53
SVM	99,39	98,18
Ensemble	**99,95**	**99,91**

Table 2. Comparison of performance measures using deep learning

Model	F1-Score (%)	Accuracy (%)
CNN	99,03	98,09
LSTM	98,78	97,59
Ensemble	**99,37**	**98,75**

- Ensemble learning turned out to be the most effective approach for malware detection.
- Individual base classifiers and the suggested ensemble methods' accuracy comparison shows that all ensemble methods outperform RF, which is the best base classifier in terms of accuracy.

Ensemble learning, while enhancing malware detection through the combined strength of multiple models, introduces several potential challenges that must be navigated carefully. One primary concern is the computational cost and complexity that escalate as the volume of data increases. Training multiple models on large datasets not only demands extensive computational resources but also requires sophisticated parallel processing and efficient data handling techniques to manage the workload effectively. Furthermore, the risk of overfitting may increase with larger datasets, especially if the diversity of the data is not adequately represented, leading to models that perform well on training data but poorly on unseen, real-world malware samples. We propose that in our future work, we should focus on large datasets, to evaluate the effectiveness of ensemble models, also to mitigate these challenges and leverage the full potential of ensemble learning in malware detection.

5 Conclusion and Future Scope

In recent years, malware has grown exponentially and gained popularity. Furthermore, malware is becoming increasingly creative and complex. As a result, it has been shown that traditional methods of malware detection are inefficient.

In this paper, we have offered an effective approach for ensemble learning-based malware detection. We have proposed two methods for an ensemble classifier. A dataset of 42 797 malware API call and 1079 goodware API call is used to assess the suggested approaches, and they are compared with individual classifiers. According to the experiment's results, the proposed method offers the highest accuracy of 99.91%.

In this work, a simple dataset was used. In the future, we tend to use larger and more complicated datasets. Further, we aim to expand the dataset's features. Additionally, our future goal is to offer a hybrid ensemble learning that may combine both machine learning and deep learning in one single ensemble.

References

1. What Is an Advanced Persistent Threat (APT)?—kaspersky.com. https://www.kaspersky.com/resource-center/definitions/advanced-persistent-threats. Accessed 25 Dec 2023
2. The Melissa Virus—Federal Bureau of Investigation—fbi.gov. https://www.fbi.gov/news/stories/melissa-virus-20th-anniversary-032519. Accessed 24 Dec 2023
3. Radware. MyDoom: An Overview of the MyDoom Virus—Radware—radware.com. https://www.radware.com/security/ddos-knowledge-center/ddospedia/mydoom/. Accessed 24 Dec 2023
4. What Is Stuxnet?—Trellix—trellix.com. https://www.trellix.com/security-awareness/ransomware/what-is-stuxnet/. Accessed 24 Dec 2023
5. What was WannaCry?—WannaCry Ransomware—Malwarebytes—malwarebytes.com. https://www.malwarebytes.com/wannacry. Accessed 24 Dec 2023
6. What is Emotet Malware? Definition, infection chain and protection!—hornetsecurity.com. https://www.hornetsecurity.com/en/knowledge-base/emotet/. Accessed 24 Dec 2023
7. CovidLock: Android Ransomware Walkthrough and Unlocking Routine—zscaler.com. https://www.zscaler.com/blogs/security-research/covidlock-android-ransomware-walkthrough-and-unlocking-routine. Accessed 24 Dec 2023
8. What are the different types of malware?—kaspersky.com. https://www.kaspersky.com/resource-center/threats/types-of-malware. Accessed 24 Dec 2023
9. Bagging, Boosting, and Stacking in Machine Learning—Baeldung on Computer Science—baeldung.com. https://www.baeldung.com/cs/bagging-boosting-stacking-ml-ensemble-models. Accessed 24 Dec 2023
10. Yerima, S.Y., Sezer, S., Muttik, I.: High accuracy android malware detection using ensemble learning. IET Inf. Secur. **9**(6), 313–320 (2015)
11. Feng, P., Ma, J., Sun, C., Xinpeng, X., Ma, Y.: A novel dynamic android malware detection system with ensemble learning. IEEE Access **6**, 30996–31011 (2018)
12. Li, S., Li, Y., Han, W., Xiaojiang, D., Guizani, M., Tian, Z.: Malicious mining code detection based on ensemble learning in cloud computing environment. Simul. Model. Pract. Theory **113**, 102391 (2021)
13. skycure malwara android android smartphones are almost twice as likely as ios devices - recherche google. https://docs.broadcom.com/doc/skycure-mobile-threat-intelligence-report-q1-2016-en. Accessed 24 Dec 2023
14. Christianah, A., Gyunka, B., Oluwatobi, A.: Optimizing android malware detection via ensemble learning (2020)
15. Idrees, F., Rajarajan, M., Conti, M., Chen, T.M., Rahulamathavan, Y.: Pindroid: a novel android malware detection system using ensemble learning methods. Comput. Secur. **68**, 36–46 (2017)
16. Gartner—Delivering Actionable, Objective Insight to Executives and Their Teams—gartner.com. https://www.gartner.com. Accessed 24 Dec 2023
17. Vasan, D., Alazab, M., Venkatraman, S., Akram, J., Qin, Z.: MTHAEL: cross-architecture IoT malware detection based on neural network advanced ensemble learning. IEEE Trans. Comput. **69**(11), 1654–1667 (2020)
18. Yan, J., Qi, Y., Rao, Q., et al.: Detecting malware with an ensemble method based on deep neural network. Secur. Commun. Netw. **2018**, 7247095 (2018)

19. Malware Analysis Datasets: API Call Sequences—kaggle.com. https://www.kaggle.com/datasets/ang3loliveira/malware-analysis-datasets-api-call-sequences. Accessed 24 Dec 2023
20. Thankachan, K.: What? When? How?: ExtraTrees Classifier—towardsdatascience.com. https://towardsdatascience.com/what-when-how-extratrees-classifier-c939f905851c. Accessed 24 Dec 2023

Enhancing Security Through Data Analysis and Visualization with ELK

Zineb Bakraouy[1(✉)], Wissam Abbass[2], Amine Baina[3], and Mostafa Bellafkih[3]

[1] Mohammed First University, ENSAO, SmartICT Lab, Oujda, Morocco
Zineb.bakraouy@gmail.com
[2] Research Laboratory in Intelligent and Sustainable Technologies (LARTID), ENSA Marrakech - Cadi Ayyad University (UCA), Marrakech, Morocco
abbass@inpt.ac.ma
[3] National Institute of Posts and Telecommunications, Rabat, Morocco
{baina,bellafkih}@inpt.ac.ma

Abstract. In a world where data is an organization's most valuable asset, the challenge lies not only in securing it but also in extracting actionable insights. This paper explores the symbiotic relationship between data analysis, visualization, and security, with a particular focus on the ELK Stack (Elasticsearch, Logstash, Kibana). By harnessing the power of ELK, businesses can elevate their security posture by effectively analyzing and visualizing data, enabling swift threat detection and informed decision-making. Our use case consists in studying and setting up a solution for managing logs and security events. This solution, which will allow the collection and the transformation of the logs of various bricks of the IS, aims in particular to obtain information allowing the detection of abnormal behavior in order to remedy it with the necessary measures, and the reduction of the response time to incidents that may partially or completely paralyze information system activities.

Keywords: ELK · Elasticsearch · Logstash · Logging · Logs · Event · Grok · Geoip · Centos · Beats · Syslog

1 Introduction

In front of incidents and various threats impacting information system security [1, 2], it is essential to inspect the logs of critical platforms [3, 4]. This serves two primary purposes: firstly, for understanding root causes and resolving issues, as well as monitoring application performance to enhance the user experience [5]; secondly, for seeking indicators of compromise and identifying potential intrusion attempts to strengthen security against cyberattacks [6]. With this objective in mind, organizations are required to collect and analyze events, logs, and operation traces to gain a better understanding of the system's behavior and to take the necessary reactive and proactive measures [7, 8] to ensure the security and sustainability of their installations [9, 10]. However, given the enormous volume of unstructured log data, operational teams are often required to make significant efforts to extract useful information from system logs using native event managers [11]. Moreover, devices such as routers, switches, and some firewall equipment lack storage space for log data retention. To address this issue, it is necessary to send this data to an external system for processing and retention. Additionally, public organizations are required to retain log data for compliance and regulatory purposes to establish a trust ecosystem [12]. However, system logs often contain a large amount of data for minimal relevant information and need to be filtered to isolate and retain only useful information. Hence, the need for a log management solution that allows for the necessary transformations. This project aims to establish a centralized system for managing system logs and security events. This solution should not only facilitate the collection, standardization, and aggregation of logs and security events but also make it easy to analyze this data to enhance security operations. We analyze the results obtained from the ELK implementation, discussing the overall health status derived from log data, the distribution of events generated by different systems, optimization of web applications based on performance analysis, geolocation of blocked connection sources, and user tracking within the integrated system. These insights demonstrate the effectiveness of ELK in providing actionable intelligence for security enhancement and performance optimization. Our paper consists in studying and setting up a solution for managing logs and security events. The paper is organized as follows: Sect. 2 introduce a challenges of log management, in Sect. 3 we present Loggin in ISO Standars, in Sect. 4 we discuss the architecture and the logs used in our Test Lab of data analysis and visualization with ELK, Sect. 5 is dedicated to present our results and Sect. 6 for discussion, eventually in Sect. 7 we draw our conclusions.

2 Challenges of Log Management and ELK Stack

Given the increasing number of incidents that threaten information system security, it is essential to monitor the logs of critical platforms to understand the sources of threats and the root causes of incidents. Organizations are, therefore, required to monitor and analyze events, logs, and traces to comprehend the system's behavior at a given time and take proactive or corrective actions as needed. Indeed, log monitoring enables IT operations teams to identify intrusion attempts and misconfigurations, track application performance to enhance the user experience, bolster security against cyberattacks, perform root cause analysis, analyze behavior, and measure system performance indicators. Implementing a centralized and intelligent log processing solution enhances operational efficiency and reinforces information system security. However, it presents several challenges, including [13, 14]:

- Logs from different solutions are not standardized.
- A large volume of log data is generated every day.
- During incidents and peak periods, the number of generated events increases exponentially.
- Some systems produce logs containing sensitive information that must be protected during transfer, processing, and storage.
- Events need to be timestamped with very high precision.
- The procurement and maintenance costs of proprietary solutions are high.

The ELK Stack is a comprehensive solution for log and data analysis. Elasticsearch efficiently stores and retrieves data, Logstash ensures data is properly processed, and Kibana provides an intuitive interface for data exploration and visualization. This stack is widely used across industries for various applications, including log management, security monitoring, and performance analysis It is open-source, scalability, and flexibility [15, 16]:

- **Elasticsearch:** Elasticsearch is the heart of the ELK Stack. It's a highly scalable and distributed search and analytics engine that's particularly adept at handling large volumes of data in real-time. Elasticsearch is known for its ability to index and search data rapidly, making it an invaluable tool for log and event data analysis. It's often used for full-text search, data exploration, and as a centralized repository for various types of data. Its distributed nature means it can be seamlessly scaled to handle massive datasets and deliver fast search results.
- **Logstash:** Logstash is the data processing engine of the ELK Stack. It's responsible for ingesting, parsing, transforming, and enriching data from different sources before it's indexed in Elasticsearch. Logstash supports a wide range of inputs, including logs, events, and streams from diverse systems, making it a versatile tool for data integration. It enables the standardization of data formats, enhances data quality, and allows for custom data enrichment through filters. Logstash acts as the bridge between data sources and Elasticsearch, ensuring that data is well-prepared for analysis.
- **Kibana:** Kibana serves as the visualization and exploration component of the ELK Stack. It provides a user-friendly web interface that allows users to interact with the data stored in Elasticsearch. With Kibana, users can create custom dashboards, conduct ad-hoc searches, and generate visualizations to gain insights from the data. It offers a variety of visualization options, including charts, graphs, maps, and tables. Kibana's dynamic and interactive dashboards make it easy to monitor data in real-time, identify trends, and troubleshoot issues. It's a powerful tool for data visualization and analysis, particularly in the context of log and event data.

3 Logging in the ISO Standard

Reformulated in English, the Moroccan ISO 27001 standard, which incorporates the content of ISO 27001, requires the establishment of secure systems for the collection, processing, and storage of logs. The Moroccan ISO 27002 standard emphasizes the importance of event logging as the foundation for automated monitoring systems capable of generating consolidated reports and security alerts.

3.1 Event Logging and Monitoring

According to the Moroccan ISO 27002 standard, it is essential to create, maintain, and regularly review event logs that record user activities, exceptions, failures, and security-related events [17]. When implementing event logging, the logs should, when relevant, contain the following information:

1) User identification information and system activities.
2) Date, time, and details of significant events, such as logins and logouts.
3) The identity or location of the terminal if possible and the system identifier.
4) Records of both successful and failed access attempts to the system.

5) Records of successful and failed attempts to access data and other resources.
6) Changes made to the system configuration.
7) Use of privileged accounts and Usage of utilities and applications.
8) Files that were accessed and the nature of the access.
9) Network addresses and protocols.
10) Alarms triggered by the access control system.
11) Activation and deactivation of protective systems, such as antivirus and intrusion detection systems.
12) Records of user transactions within applications.
13) According to ISO 27002, event logging forms the basis for automated monitoring systems capable of generating consolidated reports and security alerts.

3.2 Protection of Logged Information

Logged data often contains crucial information for assessing system health and detecting microscopic anomalies. It serves as the primary source of information used in audits and investigations in case of malicious modifications. To ensure the security of this information source, measures should be designed to protect the logging medium against unauthorized modifications of logged information and malfunctions, including:

- Alteration of recorded message types.
- Modification or deletion of log files.
- Exceeding the storage capacity of the log file storage, which can prevent the recording of events or overwrite previously recorded events.

3.3 Administrator and Operator Logs

Users with privileged accounts may attempt to manipulate logs on information processing systems they directly control to cover up errors or unauthorized actions within their duties. It is necessary to protect and regularly review logs to ensure the accountability of users with elevated privileges. According to the Moroccan ISO 27002 standard [17], it is recommended to log the activities of system administrators and system operators and protect and regularly review the logs to ensure the compliance of system and network administration activities. To minimize the risk associated with the use of high-privilege accounts, it is advisable to employ an intrusion detection system outside the control of system and network administrators.

3.4 Clock Synchronization

NTP Service Architecture: Proper clock configuration is crucial to ensure the accuracy of audit logs that may be used in investigations or as evidence in legal or disciplinary proceedings. Inaccurate audit logs can hinder investigations and undermine the credibility of evidence. According to ISO 27002 [17], in the section concerning logging systems, it is recommended to use a master clock connected to a national atomic clock signal for synchronization. The use of the NTP protocol is suggested to ensure synchronization of all servers with the master clock. Among the recommendations of ISO 27002 [17], it is important to synchronize the clocks of all relevant information processing systems in an organization or security domain to a single reference time source. In your case, you have configured your NTP reference servers to obtain time from a global NTP cluster accessible via the internet. In case of internet access issues, you have set up an internal NTP cluster composed of several servers to ensure service continuity. NTP System Documentation: According to ISO 27002, it is necessary to document internal and external requirements related to time representation, synchronization, and accuracy. These requirements may include legal, regulatory, contractual, compliance with standards, or internal monitoring requirements. The organization should establish a standard reference time. Documenting and implementing the organization's method for obtaining a reference time from one or more external sources and for reliably synchronizing internal clocks is also crucial [17].

4 Data Analysis and Visualization with ELK

4.1 Architecture of Test Bed

We have implemented a centralized architecture consisting of a central site and 50 remote sites. Connections between the remote sites and the central site are facilitated through two independent VPN networks. As for the applications, they are hosted in a data center at the central site. The overall network architecture primarily comprises two levels of firewalls. The first level contains front-end firewalls, which are connected to both Internet load balancing equipment and back-end firewalls. Multiple physically distinct DMZ networks are established around the front-end firewalls. The second level consists of back-end firewalls, which are connected to both the front-end firewalls and the Aggregating Switches. We have deployed the log management solution on servers hosted at the central site, responsible for collecting, transforming, and indexing logs primarily from the front-end firewalls, web servers, and infrastructure servers.

4.2 Logs

The sizing of the solution primarily depends on the size of logs generated in terms of both quantity and volume over defined durations based on the processing and retention requirements for these logs. This section provides statistics on the logs actually generated by the 9 systems covered by the project (Table 1).

The graph (Fig. 1) illustrates the trend in the number of events generated by the 9 typical systems we have integrated into the ELK solution. It is evident that this count exhibits variations across different days of the week and throughout the month.

The sizing of the platform that will host the components of the ELK solution is a critically important step for the project's success. This phase involves determining the hardware resources required to run the programs that make up the solution, including the amount of RAM, the number of CPUs, and disk space. While the size of RAM and the number of CPUs can be upgraded at any stage of deployment without negatively impacting data, storage space, on the other hand, must be calculated from the outset to avoid potential data loss or complex operations to extend or reduce disk size. In our case, we will primarily focus on storage sizing, while we will use the recommended values from the solution's vendor for RAM and CPU. The Elasticsearch cluster comprises

Table 1. Number of events generated by target systems

Solution	60 days	Events Per Day
Firewall Frontal	4 745 689	79 095
Workstation windows 10	110 696	1 845
Application web IIS1	23 134	386
Domain Controller	3 148 662	52 478
Server DNS windows	14 606	243
Server Web apache	656 503	10 942
Server IIS2	12 155 697	202595
Antivirus Server System Logs	64 013	1067
System center Server	688 847	11 481
Total	21 607 847	360 131

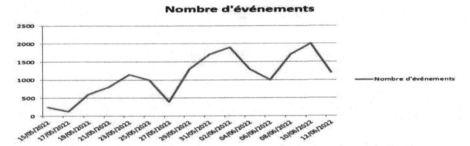

Fig. 1. Number of events per day generated by 9 typical systems

several roles, including a Master role for cluster management and a Data role for data processing. These roles can be installed on the same node or on separate nodes within the same cluster. Table 2 contains the recommended hardware prerequisites to support up to 4 GB/day [40 devices] for 30 days, depending on the cluster node configuration. To estimate the storage required for the Elasticsearch cluster, we have used the following formula, which considers the size of generated data and the storage space needed for processing by Elasticsearch, Logstash, Kibana, and the operating system itself:

Table 2. Hardware requirements to process up to 30 GB/day

Role of the node	CPU	RAM	Disk Space
Logstash and Kibana	4 Core	16 GB	40 GB HDD
Elastic Data nodes	8 Core	16 GB	200 GB SSD
Elastic Master nodes	4 Core	16 GB	120 GB SSD

We have deployed the Winlogbeat agent to retrieve logs from Windows servers and the Filebeat agent to retrieve log files from Apache and IIS servers (Fig. 2).

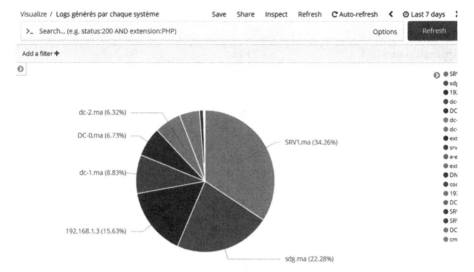

Fig. 2. Logs generated by each system

5 Results

5.1 Data Extraction with GROK

Grok is a powerful pattern-matching tool used in log management and data processing systems, particularly within the ELK (Elasticsearch, Logstash, Kibana) Stack. It enables the extraction of structured data from unstructured log files by defining custom patterns that match specific log formats. Essentially, Grok patterns help identify and extract relevant information from log entries, facilitating efficient log analysis and visualization. The GROK plugin has enabled us, among other things, to extract the data contained in the Syslog message received from the firewall. These extracted data will facilitate advanced analyses of user activity and server access (Table 3).

Table 3. Data Extraction with GROK

{ "syslog_severity": "warning", "dst_port": "3544", "received_at":"2022-05 08T21:14:30.161Z", "hashcode1": "0x0", "protocol": "udp", "src_ip": "10.1.1.55", "dst_ip": "157.56.106.184", "policy_id": "dmz_vers", "hashcode2": "0x0", "syslog_facility_code": 20, "src_port": "53591", ...}

5.2 Data Enrichment with GeoIP

After extracting the field containing the IP address with GROK, the GeoIP plugin added the following data in JSON format. This example demonstrates the power of the solution, as it allowed us to enrich log data with valuable information, including the country of origin and geographical coordinates, all from a simple IP address (Table 4).

Table 4. Data Enrichment with GeoIP

dst_geoip": { "country_code3": "US", "continent_code": "NA", "country_code2": "US", "country_name": "United States", "region_name": "Virginia", "timezone": "America/New_York", "city_name": "Boydton", "postal_code": "23917", "dma_code": 560, "location": { "lat": 36.6648, "lon": -78.3715 }

5.3 Overall Health Status

As illustrated by the histogram in bellow, the number of events collected from about ten sources during the month of May exceeds 30 million events. With such a large volume of data, we have observed that the system follows patterns that repeat based on the time of day, day of the week, or the period of the month. During attacks or incidents in general, the system's behavior changes rapidly. Monitoring log files, among other things, allows us to identify periods of unusual activity in order to react promptly in the event of a silent incident (Fig. 3).

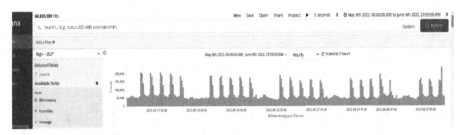

Fig. 3. State health

Systems operating in load-balanced clusters should, in principle, generate a similar quantity of logs, as depicted in the pie chart in Fig. 4 for the case of domain controllers.

5.4 Optimizing of Web Applications

Users of the integrated system frequently complain about slow web access to the SID platform. In an effort to make the most of this situation, we have configured the IIS web

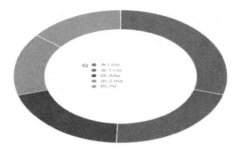

Fig. 4. Distribution of events generated by domain controllers

server hosting the application to send its logs to the ELK solution. Upon receiving the logs, we configured Logstash's GROK filters to extract the data contained within them. Subsequently, this data is sent to Elasticsearch for indexing. When numerical data is present in the logs, its utility remains limited unless it is extracted and converted into digital values for use in mathematical operations. One use case for this data involves calculating the average execution time of web pages generated by IIS, as shown in Fig. 5. This enables us to identify and address bottlenecks within a web application to optimize the platform and enhance the user experience.

Fig. 5. Performance Analysis

5.5 Geolocation of Blocked Connection Sources

Logs containing public IP addresses can be enriched with additional information using the GeoIP plugin at the Logstash level. The GeoIP plugin utilizes a local database of public IP addresses with various details such as latitude, longitude, country, and city. In our case, we configured the Logstash pipeline to extract information from logs originating from the front-end firewall in SYSLOG format. These logs are then enriched with additional details if they contain public IP addresses before being sent to Elasticsearch. Subsequently, we configured a visualization in Kibana to display the sources of blocked connections on a map, as depicted in the screenshot in Fig. 6.

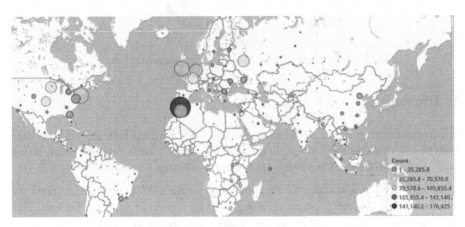

Fig. 6. Connections blocked at the front-end firewall

5.6 User Tracking Within SID

This example showcases the flexibility provided by the ELK solution for log processing, given the necessary configuration. We configured Logstash to extract information from logs originating from web servers using the GROK filter. Among this information is a string that contains details about the browser used to access SID. We pass this string to the Agent plugin, which enriches these logs with browser name and version information. The pie chart in Fig. 7 represents the list of the most commonly used web browsers to access the integrated information system SID.

This graph is the result of aggregating events from the IIS server based on the browser name over a 7-day period. So, we will delve further into the logs from IIS. Since Chrome is the most commonly used browser by SID users, we will closely examine the distribution of its versions. The pie chart in Fig. 7 displays the percentages of Chrome browser version. To produce this result, we filtered the data containing Chrome and then aggregated it based on versions.

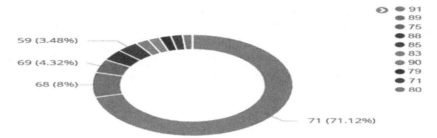

Fig. 7. Chrome browser versions of SID users over 7 days

6 Discussion

The implementation of the ELK (Elasticsearch, Logstash, and Kibana) stack for data analysis and visualization in our centralized architecture yielded insightful results across various aspects of log management and system monitoring.

- Data Extraction with GROK: The utilization of the GROK plugin facilitated the extraction of valuable data from Syslog messages received from the firewall. This extracted data not only enables advanced analyses of user activity and server access but also provides a foundation for deeper insights into network security and performance monitoring.
- Data Enrichment with GeoIP: After extracting IP addresses using GROK, the GeoIP plugin enriched log data with geographical information such as country of origin and coordinates. This enrichment significantly enhances the contextual understanding of log events, aiding in pinpointing potential security threats and optimizing network resources.
- Overall Health Status: The comprehensive analysis of log data revealed a significant volume of events collected from various sources, exceeding 30 million events during the observed period. Monitoring these logs enabled the identification of patterns in system behavior, facilitating the detection of anomalies and rapid response to security incidents or performance issues.
- Optimizing Web Applications: Integration of IIS logs into the ELK solution allowed for detailed performance analysis of web applications. By extracting and analyzing numerical data, such as execution times of web pages, bottlenecks within the application infrastructure were identified and addressed, leading to optimizations aimed at enhancing user experience.
- Geolocation of Blocked Connection Sources: Logs containing public IP addresses were enriched with geographical information using the GeoIP plugin, enabling visualization of blocked connection sources on a map. This visualization aids in identifying geographic patterns of malicious activity and strengthens network security measures.
- User Tracking within SID: Through configuration of Logstash and utilization of the Agent plugin, user tracking within the integrated information system SID was achieved. Analysis of browser usage patterns provided valuable insights into

user behavior, facilitating targeted improvements in user experience and system performance.

Overall, the results demonstrate the effectiveness of the ELK stack in providing comprehensive data analysis and visualization capabilities for log management and system monitoring. By leveraging advanced plugins and visualization tools, actionable insights were derived, enabling proactive decision-making and continuous improvement in network security and performance.

7 Conclusion

In conclusion, the implementation of ELK (Elasticsearch, Logstash, Kibana) offers a powerful solution for enhancing security through data analysis and visualization. By effectively collecting, processing, and analyzing log data, organizations can gain valuable insights into their system's behavior, detect potential threats, and proactively respond to security incidents. ELK's capabilities, such as data enrichment with GeoIP, GROK-based data extraction, and user tracking, provide essential tools for understanding and improving security postures. Moreover, ELK's flexibility and scalability make it well-suited for handling the challenges posed by a large volume of log data, enabling organizations to filter and extract meaningful information efficiently. This, in turn, leads to more informed decision-making and the ability to identify and address security vulnerabilities and performance bottlenecks. In an era where data security is of paramount importance, ELK empowers organizations to protect their information systems, comply with regulatory requirements, and maintain a high level of trust. By centralizing log management and security event analysis, ELK supports the evolution of security practices to meet the demands of the digital age. As threats continue to evolve, ELK's capabilities ensure that organizations are equipped to safeguard their systems and respond swiftly to security challenges.

References

1. Sameer, D., Swaminathan, K.: Efficient surveillance and monitoring using the ELK stack for IoT powered smart buildings. In: 2nd International Conference on Inventive Systems and Control (ICISC), India (2018)
2. Gomez, R., Wang, Q.: An investigation into Logstash configurations for real-time data processing. Int. J. Comput. Sci. **24**(3), 210–225 (2019). https://doi.org/10.9876/ijcs.2019.54321
3. Ahmed, M., Sina, A., Chowdhury, R., Ahmed, M., Rafee, M.H.: An advanced survey on cloud computing and state-of-the-art research issues. IJCSI Int. J. Comput. Sci. Issues **9**(1), 201–207 (2012)
4. Smith, J., Brown, A.: Leveraging ELK for log data analysis: a comprehensive study. J. Data Sci. Anal. **15**(2), 123–145 (2022). https://doi.org/10.9876/jdsa.2022.56789
5. Chen, L., Patel, S.: Elasticsearch: scalability and performance in large-scale data environments. In: Proceedings of the International Conference on Big Data, pp. 112–125. (2018). https://doi.org/10.54321/bigdata.2018.87654
6. Harris, M., Smith, C.: Kibana visualizations: a practical guide. In: Advancements in Data Visualization, pp. 45–68. Springer, Heidelberg (2020)

7. Kushwaha, D.S., Maurya, A.: Cloud computing-a tool for future. Int. J. Math. Comput. Res. **1**(1), 09–14 (2013)
8. Bakraouy, Z., Abbass, W., Baina, A., Bellafkih, M.: The IT infrastructure's industrialization and mastering. J. Commun. **14**(10), 884–891 (2019)
9. Bakraouy, Z., Abbass, W., Baina, A., Bellafkih, M.: MAS for services availability in cloud of things network: monitoring and reactivity. In: 12th International Conference ON Intelligent Systems: Theories and Applications (SITA), Morocco (2018)
10. Johnson, M., Patel, A.: Elasticsearch query optimizations: a comparative study. J. Inf. Process. **30**(4), 567–580 (2021). https://doi.org/10.9876/jip.2021.12345
11. Nguyen, Q., Lee, S.: Logstash configurations for streaming data: challenges and solutions. In: Proceedings of the International Conference on Data Engineering, pp. 210–223 (2019). https://doi.org/10.54321/icde.2019.87654
12. Bakraouy, Z., Abbass, W., Baina, A., Bellafkih, M.: Fuzzy multi-agent system for automatic classification and negotiation of QoS in cloud computing. J. Autom. Mob. Robot. Intell. Syst., 56–64 (2020)
13. Smith, J., Davis, R.: Integrating ELK into existing IT infrastructures. In: Advances in Log Management, pp. 112–129. Springer, Heidelberg (2018)
14. Garcia, A., Brown, L.: Log data visualization with Kibana: best practices. J. Data Sci. Anal. **25**(2), 234–248 (2017). https://doi.org/10.9876/jdsa.2017.56789
15. Wilson, K., Kim, H.: Scalability challenges in Elasticsearch: a case study. In: Proceedings of the International Conference on Big Data Analytics, pp. 112–125 (2016). https://doi.org/10.54321/icbda.2016.87654
16. Jones, A., Taylor, M.: Kibana dashboards: design principles and examples. In: Advancements in Data Visualization, pp. 45–68. Springer, Heidelberg (2020)
17. IMANOR, Document de l'IMANOR: NM ISO/IEC 27002, Rabat, pp. 48–49 (2013)

Industry 4.0 Efficiency: Predictive Maintenance with TinyML and an Incremental Model

Abdelwahed Elmoutaoukkil[1](✉) [iD], Marouane Chriss[1] [iD], Amine Khatib[1,2], and Ahmed Mouchtachi[1]

[1] Laboratory of Complex Cyber-Physical Systems (LCCPS), National School of Arts and Crafts, Casablanca, University Hassan II, Casablanca, Morocco
elmoutaoukkilabd@gmail.com
[2] LISIC, ULCO University, Calais, France

Abstract. Predictive maintenance (PdM) represents a strategic approach to the management of industrial assets that has gained significance in the era of the Fourth Industrial Revolution. In a context where operational efficiency and downtime minimization are becoming crucial imperatives, predictive maintenance stands out as an innovative solution. Instead of relying on scheduled maintenance methods based on fixed calendars, predictive maintenance leverages technological advancements, including sensors, artificial intelligence, and machine learning algorithms, to anticipate potential failures and breakdowns of industrial equipments.

However, in actual circumstances, resistance to adopting predictive maintenance persists due to a myriad of variables, as well as the high customization it requires. In this paper, we propose a new framework for PdM, which is based on TinyML technology and incremental machine learning models capable of recognizing new problems reported by those operating the machinery. Our framework, deployed on the ESP32 card and tested on a rotating machine shows a continuous improvement in system performance, demonstrating its effectiveness and adaptability.

Keywords: Predictive maintenance (PdM) · Machine Learning (ML) · TinyML · Incremental Model · Industry 4.0 · Internet of Things (IoT)

1 Introduction

In the era of Industry 4.0, technological advancements have ushered in a new paradigm for manufacturing and industrial processes. One of the pivotal components driving this transformation is Predictive Maintenance (PdM) [1]. As industries strive for increased efficiency, reduced downtime, and optimized operational processes, PdM emerges as a critical enabler for achieving these goals [2].

Predictive Maintenance involves the use of advanced analytics, machine learning, and sensor technologies to anticipate and address potential equipment failures before they occur [3]. Unlike traditional maintenance approaches that rely on fixed schedules or reactive responses, PdM leverages real-time data and predictive analytics to forecast equipment degradation and performance issues. This proactive strategy not only minimizes unplanned downtime but also enhances overall equipment effectiveness and extends the lifespan of machinery [3].

Predictive Maintenance (PdM) stands as a cornerstone in the quest for operational excellence, employing advanced analytics and machine learning to anticipate and prevent equipment failures [4]. The incorporation of TinyML, which involves deploying machine learning models on resource-constrained devices [5], and Incremental Model updates, which enable models to adapt and improve over time, introduces a new dimension to the efficiency equation within Industry 4.0.

In this paper, we have proposed the implementation of predictive maintenance on a rotating machine using TinyML technology. The use of an incremental model not only enhances the overall efficiency of operation but also contributes to its adaptability and responsiveness [6]. By deploying these models on the ESP32 microcontroller, our goal is to demonstrate the feasibility and practicality of integrating TinyML into the predictive maintenance domain, with a specific focus on rotating machines.

2 Related Works

Real-time monitoring serves as a fundamental aspect of Industry 4.0 [7], leading to the development of various systems designed to oversee currents, pressures, temperatures, and other variables within industrial plants. Advances in micro-electro-mechanical systems enable the deployment of numerous cost-effective sensors capable of wirelessly sensing, computing, and communicating to collect information for environmental and equipment monitoring [8]. These sensors are interconnected through wireless sensor networks, transmitting data to the cloud for storage or additional processing using IoT protocols and technologies [9]. Public cloud service providers often offer IoT services utilizing standard protocols for real-time storage and analytics extraction from data, allowing the utilization of historical data to predict future equipment failures.

In certain instances, the volume of data to be sent to the cloud or the latency in transmitting data to and from sensors/actuators becomes excessive [10]. With TinyML, predictive maintenance models can be executed locally on the edge device without the need for constant cloud connectivity. This ensures that maintenance predictions can still be made even in environments with limited or intermittent internet access. The ability to process data locally means that predictions can be made with low latency. Quick identification of potential equipment failures allows for faster response times, reducing downtime and preventing more severe damage. In industrial settings with resource-constrained devices, TinyML's ability to operate efficiently on devices with limited processing power and memory makes it a practical solution for predictive maintenance applications.

Developing a universal model isn't feasible for microcontrollers (MCUs) since users exhibit diverse patterns, necessitating real-time updates with field data. Nonetheless, model training demands substantial resources, impeding the advancement and practicality of TinyML on resource-limited devices [11].

Induction motors are major actuators in most industrial factories, so TinyML-based predictive maintenance of electric motors is of special importance. This state is supported by the amount of research work on this field in recent years [12].

3 Monitoring System

The following subsections present the architecture, components and software features of the proposed monitoring system.

3.1 System Architecture

In this part, we elaborate on the implementation of our model. Let $\mathcal{M}$ denote the industrial machine on which we conduct the PdM task, and $\{s_1, s_2, \cdots, s_N\}$ represent the set of many sensors through which we monitor the machine dynamics.

Figure 1 illustrates the blocks of our system. The set of sensors is directly connected to a controller capable of real-time anomaly detection. These sensors can encompass various types, including accelerometers, sound sensors, voltage sensors, and more. Specifically, accelerometers can be employed to identify unusual vibrations, possibly associated with loosely secured components or excessive loads. Sound sensors, conversely, can be used to assess the presence of sounds not typical during the normal operation of $\mathcal{M}$, potentially indicating malfunctions. Many other sensor types can also be utilized, with the final selection contingent on the type of industrial machine for which the predictive maintenance task is intended, as well as the possible points of failure.

Nevertheless, our proposition is independent of the initial data, allowing operators to continually report new behaviors of $\mathcal{M}$. In conclusion, this feature enables the system to scale, beginning with a minimal amount of data and progressively identifying new data that enrich the model with information, thereby specializing further in specific failure scenarios.

3.2 System Software

- **SLDA algorithm:**

Linear discriminant analysis (LDA) is a widely embraced method in the field of computer vision and pattern recognition due to its versatility in both dimensionality reduction and classification tasks. Numerous studies [13, 14] have demonstrated its effectiveness. LDA involves seeking a linear data transformation that optimally separates classes while reducing the dimensionality of the data [15].

Typically, implementing the LDA technique necessitates having all samples available in advance [16]. However, there are situations where the complete data set is not available and the input data is observed as a flow. In this case, the LDA feature extraction should have the ability to update the computed LDA features by observing the new samples without running the algorithm on the entire data set. For example, in many real-time situations such as online facial recognition, mobile robotics, and IoT networks, It's crucial to refresh the extracted LDA features promptly upon the availability of new observations.

Merely by observing new samples, one can employ a streaming LDA algorithm, a concept that has received considerable attention over the last two decades. Chatterjee and Roychowdhury [14] proposed an incremental self-organizing LDA algorithm to update LDA features. Demir and Ozmehmet proposed local online learning algorithms to update

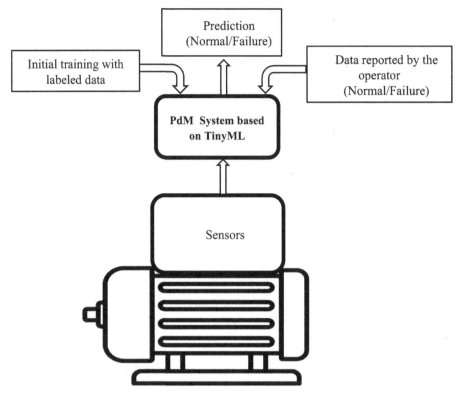

Fig. 1. The building blocks of our PdM system.

LDA features incrementally using error correction and Hebbian learning rules. Later, Aliyari et al. Derived Efficient incremental algorithms for updating LDA functionality through the observation of new samples [16].

Let be the following prediction equation:

$$y_t = W.z_t + b \tag{1}$$

Where $z_t \in \mathbb{R}^d$ is a vector,

$W \in \mathbb{R}^{k*d}$ and $b \in \mathbb{R}^k$ are updated online parameters, and k is the total number of classes.

SLDA stores one mean vector per class $\mu_k \in \mathbb{R}^d$ with an associated number $c_k \in \mathbb{R}$ and a single matrix of shared covariance $\Sigma \in \mathbb{R}^{d*d}$. When a new data point (z_t, y) arrives, the average vector and the associated counter are updated as follows:

$$\mu_{(k=y,t+1)} = \frac{c_{(k=y,t)} + z_t}{c_{(k=y,t)} + 1} \tag{2}$$

$$c_{(k=y,t+1)} = c_{(k=y,t)} + 1 \tag{3}$$

with μ as the mean of the class y has the moment t and $c_{(k=y,t)}$ is the associated counter.

For SLDA with an online variable variance, we use the following update equation:

$$\Sigma_{t+1} = \frac{t\Sigma_t + \Delta_t}{t+1} \quad (4)$$

Δ_t is calculated as follows:

$$\Delta_t = \frac{t(z_t - \mu_{(k=y,t)})(z_t - \mu_{(k=y,t)})^T}{t+1} \quad (5)$$

We use Eq. (1) and W_k to calculate the prediction.
The Columns of W is given as follows:

$$vW_k = \Lambda \mu_k \quad (6)$$

with Λ are determined by:

$$\Lambda = [(1-\varepsilon)\Sigma + \varepsilon.I]^{-1} \quad (7)$$

b_k is updated as follows:

$$b_k = -\frac{1}{2}(\mu_k.\Lambda\mu_k) \quad (8)$$

- **The proposed strategy**

Running SLDA inferences on microcontrollers is akin to processing data continuously one after another. The algorithm can handle the next input sample after the completion of the last inference. SLDA leverages this characteristic and further extends the existing model with post-training capability.

The central component of SLDA is the parameter retrieval and update phase. These parameters can be developed in real time. After deploying the system on other existing models, the models improve and adapt to new contexts. Since the initial models are downloaded as C arrays into the microcontroller's EEPROM memory, they can be modified later.

As shown in Fig. 2, when the operator observes a system misprediction, they can intervene to correct it by indicating the correct situation. The algorithm retrieves the old parameters, calculates new ones, stores them for future processing, and then proceeds to the prediction process.

- **Performance of the ML model**

a) To measure how "accurate" or superior your classifiers are at predicting the class label of tuples. The classifier evaluation measures are accuracy, sensitivity (or recall), specificity, precision, and F1. The accuracy of a classifier on a given test set is the percentage of true positives and true negatives from all correct classifiers.

$$Accuracy = \frac{TP + TN}{TP + TN + FP + FN} \quad (9)$$

Here, TP = True Positive, TN = True Negative, FP = False Positive, FN = False Negative.

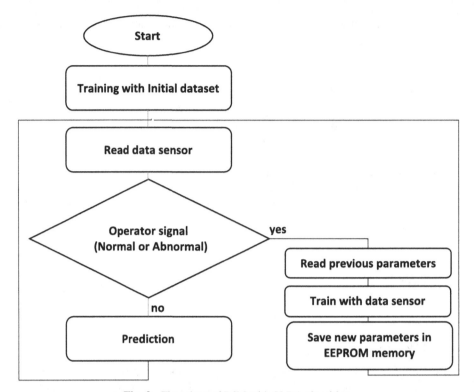

Fig. 2. Flowchart of PdM with SLDA algorithm.

4 The Experimental Plan

To assess the aforementioned concept, we conducted experiments utilizing an Mbits board [17] and an industrial fanAs depicted in Fig. 3, the board is equipped with an Xtensa working at 240 MHz, with 8 MB RAM, The specifications of the board fall within the typical type of MCUs. Additionally, it is equipped with several built-in sensors, among which the 3-axis accelerometer sensor is employed in the experiment. The fan is utilized to emulate the operation of an industrial rotating machine. More precisely, the board is employed to capture two distinct vibration patterns from the fan: normal and abnormal operation.

Figure 4 displays the data vibration of the 3-axis for each category over 40 time steps, equivalent to approximately 30 s, utilizing the accelerometer with a sampling rate of 30 Hz. The board will incorporate SLDA into the pre-trained model and classify the fan's operating mode in real-time.

The experimental setup is structured as follows:

1) Collect data of the fan from two categories: normal and abnormal operation. For each class, one minute of data is collected.
2) Apply data processing to arrange data.
3) Generate and deploy the initial model.

4) Validate the concept of continuous learning.

Fig. 3. The hardware: The Mbits board and the industrial fan.

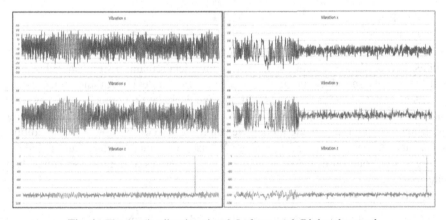

Fig. 4. The 3-axis vibration signal. Left: normal, Right: abnormal.

5 Results and Discussion

Figure 5 illustrates the evolution of the accuracy of a machine-learning model across 10 training batches. The initial accuracy stands at 69% after the initial training. Subsequently, as the model is exposed to more training data and learning iterations, its accuracy gradually improves throughout these batches. It becomes evident that the model undergoes significant improvement.

The most remarkable aspect of this training process is the substantial leap in accuracy, culminating in an impressive rate of 74% after the last (the 10th) batch. This indicates that the model has made significant progress in its ability to make accurate predictions as it learns from the data.

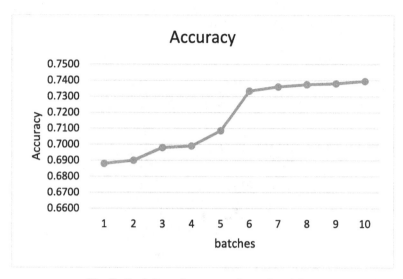

Fig. 5. Evolution of accuracy after each training.

6 Conclusion

TinyML is poised to play a crucial role in enabling large-scale machine learning applications. Yet, on-device training using microcontrollers remains a significant challenge. In this study, we introduce an innovative approach to on-device training by harnessing incremental learning. Our research highlights the degradation in model accuracy when training and inference are performed on distinct datasets. By integrating our proposed algorithm into any pre-existing model, we can tailor it to evolving working conditions and enhance its performance progressively. We have validated the effectiveness and feasibility of our solution using an Mbits card that integrates a vibration sensor with an industrial fan.

Our future work includes:

- Comparative studies between the results provided by the model generated based on raw data from the sensor and that of a model trained after feature extraction.
- Addressing the training time in streaming and prediction time.
- Analyzing the possibility of integrating real-time dimensionality reduction.
- Studying the feasibility and effectiveness of applying this approach to machines of the same category connected to the cloud, allowing each machine to share its experience with others.
- Exploring learning with reduced numerical precision.

References

1. Jan, Z., et al.: Artificial intelligence for industry 4.0: systematic review of applications, challenges, and opportunities. Expert Syst. Appl. **216** (2023)
2. Hurtado, J., Salvati, D., Semola, R., Bosio, M., Lomonaco, V.: Continual learning for predictive maintenance: overview and challenges. Intell. Syst. Appl. **16** (2023)
3. Arafat, M.Y., Hossain, M.J., Alam, Md.M.: Machine learning scopes on microgrid predictive maintenance: potential frameworks, challenges, and prospects. Renew. Sustain. Energy Rev. **190** (2024)
4. Kashpruk, N., Piskor-Ignatowicz, C., Baranowski, J.: Time series prediction in Industry 4.0: a comprehensive review and prospects for future advancements. Appl. Sci. (2023)
5. Sudharsan, B., Breslin, J.G., Intizar Ali, M.: Edge2Train: a framework to train machine learning models (SVMs) on resource-constrained IoT edge devices. In: IoT 2020, Sweden (2020)
6. Bachinger, F., Kronberger, G., Affenzeller, M.: Continuous improvement and adaptation of predictive models in smart manufacturing and model management. IET Collab. Intell. Manuf. **3**, 48–63 (2021)
7. Liu, Y., Xu, X.: Industry 4.0 and cloud manufacturing: a comparative analysis. Manuf. Sci. Eng. **139** (2016)
8. Gongora, V.C., Hancke, G.P.: Industrial wireless sensor networks: challenges, design principles, and technical approaches. IEEE Trans. Ind. Electron. **56**, 4258–4265 (2009)
9. Xu, L.D., He, W., Li, S.: Internet of Things in industries: a survey. IEEE RFID Virtual J. **10**(4), 2233–2243 (2014)
10. Magadan, L., Suárez, F., Granda, J., Garcia, F.D.: Real-time monitoring of electric motors for detection of operating anomalies and predictive maintenance. IEEE RFID Virtual J. **323**, 2233–2243 (2014)
11. Ren, H., Anicic, D., Runkler, T.A.: TinyOL: TinyML with online-learning on microcontrollers. In: 2021 International Joint Conference on Neural Networks (IJCNN), Shenzhen, China (2021)
12. Ajitha, A., et al.: IoT platform for condition monitoring of industrial motors (2017)
13. Jin, L., Ding, K., Huang, Z.: Incremental learning of LDA model for Chinese writer adaptation. Neurocomputing **73** (2010)
14. Demir, G.K., Ozmehmet, K.: Online local learning algorithms for linear discriminant analysis. Pattern Recognit. Lett. **26**(4), 421–431 (2005)
15. Abrishami Moghaddam, H., Zadeh, Kh.A.: Fast adaptive algorithms and networks for class-separability features. Pattern Recognit. 1695–1702 (2003)
16. Aliyari Ghassabeh, Y., Rudzicz, F., Moghaddam, H.A.: Fast incremental LDA feature extraction. Pattern Recognit. 1999–2012 (2015)
17. elecrow, "elecrow," (2022). https://www.elecrow.com/wiki/index.php?title=Mbits

Tiny-ML and IoT Based Early Covid19 Detection Wearable System

Oussama Elallam [1(✉)], Oussama Jami[2], and Mohamed Zaki[3]

[1] Complex Cyber-Physical Systems Laboratory, The National Superior School of Arts and Crafts, University Hassan II, Casablanca, Morocco
elallamoussama7@gmail.com
[2] High School of Technology in Salé; Materials, Energy and Acoustics Team, Mohammed V University in Rabat, Rabat, Morocco
[3] Biomedical Engineering Department, Mohammed VI University of Health Sciences, Casablanca, Morocco

Abstract. The COVID-19 pandemic has unleashed significant global challenges, straining healthcare systems and impacting millions. Traditional diagnostic methods, such as PCR tests and chest X-ray scans, while effective, face accessibility and scalability issues, particularly in developing regions. Furthermore, these methods do not offer continuous monitoring, critical for early detection and intervention to prevent severe disease progression. Addressing these gaps, our research introduces an innovative AI and IoT-based wearable health monitoring system. Designed for affordability and ease of use, this wearable device integrates vital signs sensors—heart rate, blood oxygen saturation, and body temperature—alongside a small touch screen for symptom reporting. A distinctive aspect of our device is its embedded TinyML model, utilizing an XGBoost algorithm, that facilitates immediate on-device COVID-19 infection prediction. This feature significantly enhances the device's utility by enabling real-time health status updates and early warning signals for potential COVID-19 infection, which is particularly vital in settings lacking robust healthcare infrastructure. Moreover, by eliminating the need for external data processing, our approach not only ensures data privacy and security but also addresses the critical delay in diagnosis, offering a scalable solution adaptable across various healthcare scenarios. Through this system, we aim to democratize access to COVID-19 screening and monitoring, presenting a significant advancement in wearable healthcare technologies and their application in pandemic response and beyond.

Keywords: Covid19 · Tiny Machine learning · IoT · wearable medical device

1 Introduction

The Coronavirus disease 2019 (COVID-19) pandemic, caused by the severe acute respiratory syndrome coronavirus 2 (SARS-CoV-2), has precipitated an unprecedented global crisis, straining healthcare systems and economies worldwide. Initially identified in Wuhan, China, in late 2019, COVID-19 rapidly transcended geographical boundaries, evolving into a pandemic as declared by the World Health Organization (WHO) on March 11, 2020. Despite the development and dissemination of vaccines, the battle against COVID-19 continues to face multifaceted challenges, including the emergence of highly transmissible viral variants [1, 2], varied efficacy and acceptance rates of vaccines [3], logistical hurdles in vaccine distribution [4], and concerns over the potential waning of vaccine-induced immunity over time [5].

Predominantly, the detection of COVID-19 has relied on molecular diagnostics, such as the polymerase chain reaction (PCR) technique, necessitating specialized laboratories and personnel for sample collection. While accurate, these methods face logistical challenges, including the requirement for sophisticated lab infrastructure, prolonged result turnaround times, and the inability to continuously monitor patients. Furthermore, many COVID-19 infections are asymptomatic or exhibit non-specific symptoms like fever, cough, and shortness of breath, complicating early detection efforts [6–11]. In severe cases, respiratory distress, a critical symptom, signifies a heightened risk of mortality [12]. Notably, an infected individual's blood oxygen level, typically ranging between 92% and 96% in COVID-19 cases—as opposed to the 95% to 100% range observed in healthy adults—serves as a crucial indicator of infection [13–15].

The limitations of traditional diagnostic methods and the need for continuous monitoring highlight the potential of wearable sensor technology for the non-invasive monitoring of vital signs indicative of COVID-19. The advent of wearable medical devices, augmented by the Internet of Things (IoT) and machine learning, has broadened the horizon for early disease detection and monitoring, offering a glimpse into future healthcare innovations [8, 16–22]. These technological advancements underscore the potential of wearable sensors in the early diagnosis and monitoring of COVID-19, particularly in underserved and remote areas.

In light of these considerations, this research proposes a novel wearable medical device leveraging TinyML technology for the on-device analysis of vital signs to predict COVID-19 infection. By integrating direct patient interaction through a user-friendly interface and combining it with real-time health data analysis, this device aims to overcome the limitations of current diagnostic methods and contribute to the global effort against the pandemic.

2 Research Method

Our research bifurcates the development of a pioneering wearable system into two critical undertakings: the creation of a compact, cost-efficient wearable device endowed with IoT capabilities that accurately measures heart rate, blood oxygen saturation, and body temperature; and the deployment of a refined machine learning model within a device constrained by limited RAM, while ensuring data accuracy and precision. This endeavor is further complicated by the selection of an appropriate dataset for model training. Our selection of vital signs—heart rate, blood oxygen saturation, and body temperature—as primary indicators for COVID-19 infection is grounded in their established clinical relevance for assessing patient health. These parameters are among the most commonly checked in clinical practice, offering a reliable basis for early detection of potential health issues, including but not limited to COVID-19. The universality of these parameters in health assessment allows our device to seamlessly integrate into existing healthcare monitoring practices, thereby enhancing its adaptability for both pandemic response and routine health monitoring.

The COVID-19 pandemic, which emerged in Wuhan in 2019, is characterized by a plethora of symptoms including fever, respiratory difficulty, cough, sore throat, muscle pains, headaches, and loss or alteration of taste and smell. A 2020 data analysis from approximately 25,000 adult cases revealed fever as the predominant symptom, occurring in roughly 78% of confirmed cases. The variability of these symptoms, ranging from respiratory issues to neurological effects, underscores the disease's complexity.

Given the fluctuating nature of early COVID-19 symptoms, consistent monitoring of key health indicators such as temperature and oxygen saturation is advised. Typically, oxygen saturation should fall between 95 to 99%, and normal body temperature should range from 36.5 °C to 37.5 °C. Deviations from these norms could indicate potential infections, including COVID-19. However, the lack of regular medical check-ups and the unreliability of manual readings necessitate a more dependable method of monitoring these vital signs.

To address this need, we engineered a wearable device capable of continual measurement of heart rate, blood oxygen saturation, and body temperature, transmitting this data via the internet for remote health monitoring. The device's interface, a compact touch screen, not only exhibits real-time data but also engages users for additional symptom reporting. This dual-modality data capture enhances the system's diagnostic capabilities, facilitating early detection of COVID-19.

For disease prediction, we have constructed a machine learning model, informed by a substantial dataset of COVID-19 patient symptoms, to be deployed directly onto the wearable device. Utilizing TinyML tools, this model aims to reduce the time and cost associated with COVID-19 detection and sidestep the drawbacks associated with conventional machine learning implementations that depend on cloud connectivity.

The architecture of the proposed device is centered on the ESP32 module, an MCU celebrated for its dual-core processing and integral IoT and Bluetooth functionalities. The incorporation of two primary sensors, the MAX30102 for cardiac and oximetric readings, and a 10K NTC thermistor for temperature measurements, exemplifies the device's comprehensive sensing abilities. The complexity of the machine learning code, coupled with the sensitivity requirements of the MAX30102 sensor, warranted the integration of an Arduino Mini Pro board, tasked with data acquisition and relayed via UART to the ESP32. Data visualization and user interaction are mediated through an ILI9341 touch screen. The device's power supply is managed through an accompanying circuit, ensuring sustained operation, and is housed within an ergonomically designed tablet bracelet, maximizing user comfort. Figure 1 illustrates the block diagram of our system, detailing the interconnections and functional flow.

Upon data acquisition, the wearable device not only dispatches the information to cloud servers but also conducts a 24-h drift analysis to pinpoint subtle yet critical physiological changes. Should such changes be detected, the device prompts the user with a survey to garner additional information pertinent to their travel history, possible family exposure to COVID-19, and symptoms untraceable by sensors, like muscle pain and

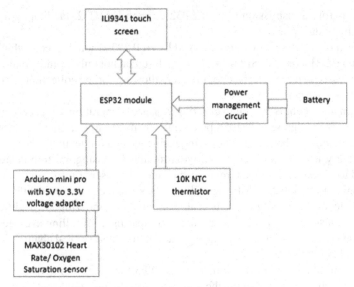

Fig. 1. Block diagram of the proposed system

fatigue. The collated data, encompassing sensor readings and user responses, feeds into the pre-deployed machine learning algorithm on the device. This model was meticulously trained using a dataset curated by Beijing University's Big Data High-accuracy Center, which was methodically sourced from governmental COVID-19 tracking websites. A collaboration with other researchers led to the distillation of key dataset features—age, gender, fever, cough, muscle soreness, lung infection, radiographic findings, pneumonia, diarrhea, runny nose, isolation status, and travel history—all pertinent to COVID-19 symptomatology [23].

2.1 Hardware Design

The hardware foundation of our wearable diagnostic system is outlined in Fig. 2, demonstrating an orchestrated network of components that ensure multifaceted health monitoring. Central to this network is the ESP32 module, esteemed for its robust processing capabilities and IoT functionalities. It serves as the primary data processor and communication hub within our system.

Temperature measurement is crucial for accurate COVID-19 detection. Although the MAX30102 sensor includes a built-in temperature sensor, its primary contact point is the finger, which does not yield reliable body temperature readings critical for our analysis. Consequently, we have integrated a 10K NTC thermistor, interfaced with the ESP32's analog pin 33, to ensure precise and reliable core body temperature measurements, essential for our device's predictive accuracy.

For the critical tasks of heart rate and blood oxygen saturation measurement, we employ the MAX30102 sensor, which communicates with an Arduino Mini Pro via an I2C connection. This Arduino Mini Pro acts as an intermediary to preprocess the sensor data before transmission to the ESP32 through UART 2, facilitating a modular and streamlined data pipeline.

Display and user interaction are managed by an ILI9341 touchscreen, which is connected to the ESP32 through SPI pins. This touch screen not only exhibits real-time data but also serves as an input interface for the user, allowing for the collection of subjective health metrics via a survey interface.

Power management is pivotal to ensuring consistent operation, especially in a wearable context. We integrate a TP4056 power management circuit, drawing energy from a rechargeable lithium battery, thus offering resilience and longevity to the system. The entire assembly is encased in compact, wrist-mounted housing, custom-designed and 3D-printed to accommodate the circuitry while ensuring user comfort and ergonomics.

Leveraging the ESP32's Wi-Fi capabilities, the system can transmit collected health data to the cloud, facilitating not only real-time remote monitoring for healthcare professionals but also historical data analysis. By comparing 24-h rolling averages against live data, the system can detect and flag any anomalous physiological changes indicative of a potential health issue.

Upon identification of such anomalies, the ESP32 initiates an interactive questionnaire on the touch screen, prompting the wearer to provide additional contextual information pertinent to their symptoms and exposure history. This human input data, coupled with sensor-derived biometrics, is then used to populate a feature set for subsequent analysis.

A pre-trained machine learning model residing on the ESP32 evaluates this feature set to assess the likelihood of a COVID-19 infection. In the event of a high-risk assessment, the system alerts the wearer with recommendations for further testing and precautionary measures. Simultaneously, an automated alert, including the assessment results, is dispatched to the wearer's designated healthcare provider to expedite professional intervention.

2.2 Software Development

The software infrastructure forms the intellectual backbone of our wearable system, underpinning the predictive capabilities of our device. The ESP32 module, serving as the central processor, hosts a machine learning model that we meticulously engineered using the versatile Python programming language. This choice was motivated by Python's extensive library ecosystem and its dominance in the data science domain.

We entrusted the task of training, validating, and testing our machine learning model to this robust language, exploiting libraries like XGBoost for model development and pandas along with NumPy for data manipulation. The critical phase of transitioning our machine learning model from a Python-based prototype to a deployable entity on the ESP32 was achieved through the translation of the model into C code. This transpilation process was facilitated by the m2cgen library, chosen for its capability to convert complex models into native code without imposing additional dependencies, allowing for lean and efficient on-device execution.

The dataset, previously subjected to rigorous preprocessing by researchers in [23], forms the empirical bedrock upon which our predictive model is constructed. It consists of a balanced compilation of symptomatic features, both physiological and subjective,

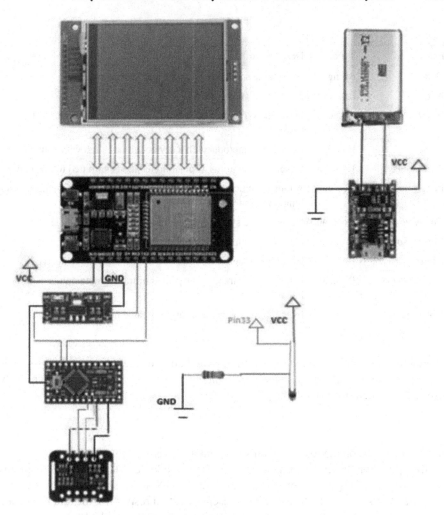

Fig. 2. Hardware architecture

carefully selected for their association with COVID-19. The dataset encapsulates confirmed and suspected cases, resulting in a rich matrix of features detailed in Table 1, including gender, age, and a range of symptomatic indicators.

A comparative analysis, as detailed in the study [23], benchmarked various machine learning algorithms to ascertain their predictive prowess. The XGBoost model emerged as the paragon, distinguished by its precision, recall, F1 scores, and AUC values, which collectively attest to its accuracy in diagnosing COVID-19.

The XGBoost's objective function is formulated as:

$$Obj = \sum_{i=1}^{n} l(y_i, \hat{y}_i) + \sum_{k=1}^{K} \Omega(f_k), \tag{1}$$

Table 1. Features description.

Feature	Type	Description
Gender	String	Almost the same ratio of male and female patients
Age	Integer	The age range is 0–96 years
Fever	Boolean	Develops symptoms with a high body temperature of 38 °C or more
Cough	Boolean	Develops symptoms with a dry cough or cough with sputum
Pneumonia	Boolean	Develops symptoms of pneumonia and is admitted to hospital
Lung infection	Boolean	A radiographic or CT scan indicates chest imaging changes as a lung infection
Runny nose	Boolean	Develops the symptom of a runny nose
Muscle soreness	Boolean	Develops symptoms of limb or muscle soreness
Diarrhea	Boolean	Develops symptoms of diarrhea and is admitted to hospital
Travel history	Boolean	Patients are marked as suspected of travelling to one or more tracks
Isolation	Boolean	Isolation treatment status at designated hospitals

where l denotes the loss function quantifying prediction discrepancies, and Ω signifies the regularization component that mitigates model overfitting. The algorithm iteratively refines its predictions through:

$$\hat{y}_i^{(t)} = \hat{y}_i^{(t-1)} + f_t(x_i), \qquad (2)$$

optimizing based on the loss function's gradient and Hessian (Fig. 3).

In application, upon detecting aberrations in physiological data, the ESP32 springs into action, gathering a suite of 11 features to feed the XGBoost model. These features are a blend of sensor-derived metrics and user-submitted responses, the latter obtained through a succinct questionnaire on the device's touch screen (Table 2).

The ESP32 will calculate abnormal differences between the current data and the mean. Those abnormal changes can be identified as follows: for the heart rate, according to [24], if the average of 24H is above 90 bpm and the current heart rate is between 90–100 bpm, then it might be a sign of tachycardia caused by infection. For the blood oxygen saturation, if a decrease in oxygen saturation of >3% from baseline (which is the average of 24H) and the current oxygen saturation is <94%, then it's highly likely that the patient has pneumonia which is a symptom of covid19 disease [25]. For the body temperature, if the average of 24H is above 37.2 °C (>99 °F) and the current body temperature is >37.7 °C (>100 °F), then the patient has a fever which is a primary symptom of COVID-19.

Should the model identify a high likelihood of COVID-19 infection, the device issues an immediate notification to the user, urging confirmatory PCR testing and self-isolation. Concurrently, it triggers an automatic email alert to the user's healthcare provider, potentially curtailing further transmission of the virus.

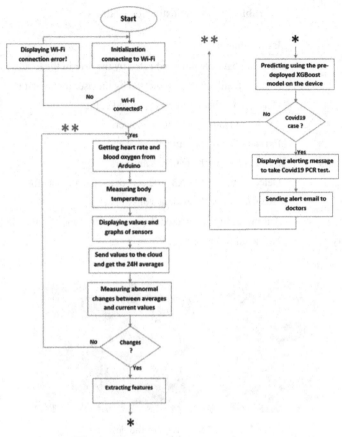

Fig. 3. Flowchart of the proposed system

3 Results and Discussion

The primary interface of our wearable device presents critical health metrics including heart rate, blood oxygen saturation (SpO2), body temperature, and the perfusion index. The perfusion index offers insights into pulse strength, enhancing the device's utility in monitoring vital signs crucial for early COVID-19 detection. This information is succinctly displayed on the main screen for ease of user interpretation (refer to Fig. 4).

In alignment with modern IoT paradigms, our system seamlessly transmits these health metrics to the cloud in real-time via Thingspeak, a robust IoT platform sponsored by MathWorks. This feature facilitates remote monitoring by healthcare professionals and supports data analysis for continuous improvement of predictive algorithms (illustrated in Fig. 5).

Upon detecting significant variations in the monitored health parameters, our device initiates an adaptive survey comprising six binary (yes/no) questions. These inquiries are designed to refine the dataset with patient-specific symptoms and exposures, thereby augmenting the accuracy of COVID-19 prediction (as depicted in Figs. 6 and 7).

Table 2. Features extraction methods.

Feature	Extraction method
Gender	A static feature will be indicated by the user in the first use
Age	A static feature will be indicated by the user in the first use
Fever	Detected by thermistor as explained before
Cough	Extracted from the patient's answers
Runny nose	Extracted from the patient's answers
Muscle soreness	Extracted from the patient's answers
Pneumonia	Detected by the MAX30102 sensor as explained before
Diarrhea	Extracted from the patient's answers
Lung infection	Automatically considered if there are fever and pneumonia [26]
Travel history	Extracted from the patient's answers

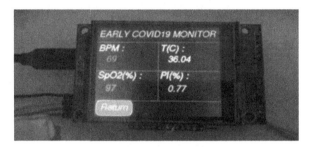

Fig. 4. Main screen of the device

After survey completion, the device promptly evaluates the collected data to predict the likelihood of COVID-19 infection. Positive predictions are immediately communicated to the user via the device's display (Fig. 8). For cases deemed high-risk, the device activates an onboard buzzer as an immediate alert and dispatches an automated email to designated healthcare providers. Concurrently, it delivers precautionary advice to the user to mitigate virus transmission.

The TinyML model's efficacy was thoroughly evaluated through key performance metrics validating its diagnostic precision. It achieved a Precision of 0.89–0.93, indicating high true positive identification reliability. Recall was 0.90–0.94, demonstrating effectiveness in correctly identifying actual positives. The F1 Score of 0.88–0.93 showcased balanced precision and recall accuracy. Additionally, the AUC of 0.82–0.87 signified excellent COVID-19 patient discrimination capability. Cumulatively, these metrics highlighted the XGBoost model's robustness in accurate COVID-19 prediction, justifying its integration into the device.

Fig. 5. Data visualization in the cloud (Thingspeak platform)

Fig. 6. First page of the survey   **Fig. 7.** Second page of the survey

Discussion

This study validates the feasibility of integrating advanced sensor technology with real-time data analysis for early COVID-19 detection. The device's continuous monitoring and adaptive response system underscores its potential as a pivotal tool in pandemic management. Moreover, the utilization of cloud-based data aggregation enhances the scalability of this approach, allowing for real-time data access by healthcare professionals and facilitating a proactive healthcare response.

To enhance the versatility and applicability of our wearable device, it is designed to serve not only as an early detection tool for COVID-19 in vulnerable populations and during pandemics but also as a robust system for the routine remote monitoring of patients' vital signs. This dual functionality underscores its utility in both crisis scenarios and standard healthcare settings. By focusing on universally crucial health parameters such as heart rate, oxygen saturation, and body temperature, our device aligns with

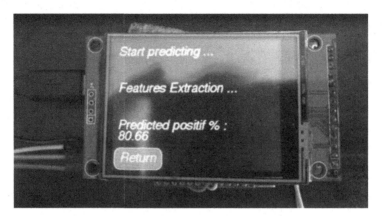

Fig. 8. Predicting of Covid19 disease

common clinical practices for assessing patient health, making it adaptable across various healthcare scenarios. Future iterations will explore customization options and additional sensor integrations to accommodate a broader spectrum of healthcare needs, emphasizing scalability and adaptability in diverse environments.

However, we acknowledge the limitations associated with relying solely on a set of physiological parameters and self-reported symptoms for COVID-19 diagnosis. While these indicators are pivotal, the specificity of COVID-19 symptoms can overlap with other respiratory conditions, necessitating further empirical validation to refine the predictive model's accuracy.

Future iterations will explore the integration of additional biometric sensors and the application of more sophisticated machine learning algorithms to enhance diagnostic precision. Moreover, scalability and the device's adaptability across diverse healthcare settings remain paramount for broader application.

4 Conclusion

This study demonstrates the potential of Tiny Machine Learning (TinyML) for early disease detection, showcasing a novel wearable device for continuous health monitoring that combines IoT with TinyML. Our device's integration of an XGBoost model for real-time COVID-19 prediction, based on physiological measurements and patient survey responses, marks a significant innovation. This approach not only enhances diagnostic accuracy but also ensures data privacy and speed by enabling on-device processing without the need for external servers.

Acknowledging the current model's limitations, future work will focus on expanding the range of health parameters analyzed and undergoing rigorous validation in real-world settings. The next development phase aims at creating a self-sufficient device capable of evolving its diagnostic capabilities through continuous learning, improving both precision and adaptability.

By advancing TinyML in wearable devices, our research paves the way for smarter, proactive health management tools. This progress signifies a step forward in making advanced diagnostics more accessible, highlighting the importance of technology in enhancing patient care and disease management.

References

1. Leung, K., Shum, M.H., Leung, G.M., Lam, T.T., Wu, J.T.: Early transmissibility assessment of the N501Y mutant strains of SARS-CoV-2 in the United Kingdom, October to November 2020. Eurosurveillance **26**(1), 2002106 (2021)
2. Munitz, A., Yechezkel, M., Dickstein, Y., Yamin, D., Gerlic, M.: The rise of SARS-CoV-2 variant B.1.1.7 in Israel intensifies the role of surveillance and vaccination in elderly. medRxiv (2021)
3. Polack, F.P., et al.: Safety and efficacy of the BNT162b2 mRNA Covid-19 vaccine. N. Engl. J. Med. (2020)
4. Pagliusi, S., et al.: Emerging manufacturers engagements in the COVID-19 vaccine research, development and supply. Vaccine **38**(34), 5418–5423 (2020)
5. Anderson, R.M., Vegvari, C., Truscott, J., Collyer, B.S.: Challenges in creating herd immunity to SARS-CoV-2 infection by mass vaccination. Lancet **396**(10263), 1614–1616 (2020)
6. NSW Health, NSW Health Pathology – diagnostic COVID-19 testing. https://www.health.nsw.gov.au/Infectious/covid-19/Pages/pathology.aspx#:~:text=NSW%20Health%20Pathology%20continues%20to,health%20staff%20as%20a%20priority. Accessed May 2022
7. WorldoMeters, Coronavirus cases update (live). https://www.worldometers.info/coronavirus/. Accessed Apr 2021
8. Mirjalali, S., Peng, S., Fang, Z., Wang, C.H., Wu, S.: Wearable sensors for remote health monitoring: potential applications for early diagnosis of Covid-19. Adv. Mater. Technol. **7**(1), 2100545 (2022)
9. Qin, C., et al.: Dysregulation of immune response in patients with coronavirus 2019 (COVID-19) in Wuhan, China. Clin. Infect. Dis. **71**(15), 762–768 (2020)
10. Du, Y., et al.: Clinical features of 85 fatal cases of COVID-19 from Wuhan. A retrospective observational study. Am. J. Respir. Crit. Care Med. **201**(11), 1372–1379 (2020)
11. Bhatraju, P.K., et al.: Covid-19 in critically ill patients in the Seattle region—case series. N. Engl. J. Med. **382**(21), 2012–2022 (2020)
12. Chen, L., Dubrawski, A., Clermont, G., Hravnak, M., Pinsky, M.R.: Modelling risk of cardiorespiratory Instability as a heterogeneous process. In: AMIA Annual Symposium Proceedings, vol. 2015, p. 1841. American Medical Informatics Association (2015)
13. Shenoy, N., Luchtel, R., Gulani, P.: Considerations for target oxygen saturation in COVID-19 patients: are we under-shooting? BMC Med. **18**(1), 1–6 (2020)
14. Ali, M.M., Haxha, S., Alam, M.M., Nwibor, C., Sakel, M.: Design of Internet of Things (IoT) and android based low cost health monitoring embedded system wearable sensor for measuring SpO2, heart rate and body temperature simultaneously. Wirel. Pers. Commun. **111**(4), 2449–2463 (2020)
15. Lathiya, N., Ruqaya, P.R., Rathore, P.: Exercise induced changes in the levels of oxygen saturation among adult males and females. JLUMHS-J. Liaquat Univ. Med. Health Sci. **15**(4), 199–202 (2016)
16. Kumar, V.S., Krishnamoorthi, C.: Development of electrical transduction based wearable tactile sensors for human vital signs monitor: fundamentals, methodologies and applications. Sens. Actuators A **321**, 112582 (2021)
17. Jeong, H., Rogers, J.A., Xu, S.: Continuous on-body sensing for the COVID-19 pandemic: gaps and opportunities. Sci. Adv. **6**(36) (2020)

18. Alam, T., Qamar, S.: Coronavirus disease (Covid-19): reviews, applications, and current status. Jurnal Informatika Universitas Pamulang **5**(3), 213 (2020)
19. Nasajpour, M., Pouriyeh, S., Parizi, R.M., et al.: Internet of Things for current COVID-19 and future pandemics: an exploratory study. J. Healthc. Inform. Res. **4**, 325–364 (2020)
20. Seshadri, D.R., et al.: Wearable sensors for COVID-19: a call to action to harness our digital infrastructure for remote patient monitoring and virtual assessments. Front. Digit. Health **8** (2020)
21. Ding, X., et al.: Wearable sensing and telehealth technology with potential applications in the coronavirus pandemic. IEEE Rev. Biomed. Eng. **14**, 48–70 (2021)
22. Mohamed, H., Mohammed, R.: Data classification by SAC "scout ants for clustering" algorithm. J. Theor. Appl. Inf. Technol. **55**(1), 66–73 (2013)
23. Ahamad, M.M., et al.: A machine learning model to identify early stage symptoms of SARS-Cov-2 infected patients. Expert syst. Appl. **160**, 113661 (2020)
24. Aranyó, J., et al.: Inappropriate sinus tachycardia in post-COVID-19 syndrome. Sci. Rep. **12**(1), 1–9 (2022)
25. Kaye, K.S., Stalam, M., Shershen, W.E., Kaye, D.: Utility of pulse oximetry in diagnosing pneumonia in nursing home residents. Am. J. Med. Sci. **324**(5), 237–242 (2002)
26. Prabhu, F.R., Sikes, A.R., Sulapas, I.: Pulmonary infections. Fam. Med. Princ. Pract., 1083–1101 (2016)

Author Index

A

Abbad, Khalid I-17
Abbass, Wissam I-246
Abbes, Ali Ben II-229
Abou Ali, Mohamad I-121
Ait Elmahjoub, Abdelhafid II-145, II-238
Ait ElMahjoub, Abdelhafid II-93
Alfathi, Najlae II-131
Ali, Hussein I-121
Allali, Hakim II-173
Al-Obeidat, Feras II-229
AlTaee, May II-229
Amazou, Youssra I-93
Amenchar, Ouiam II-43
Ameur, Soufiane II-70
Ammor, Hassan I-218
Amouri, Ali II-30
Arganda-Carreras, Ignacio I-121
Ayache, Meryeme I-233
Azmani, Abdellah I-31, I-93, I-139, I-175
Azmani, Monir I-93
Azri, Abdelghani II-173

B

Baali, Sara I-74
Baghouri, Mostafa II-43
Bahatti, Lhoussain II-120
Baina, Amine I-246
Bakraouy, Zineb I-246
Bearee, Richard II-70
Beghdadi, Ayman II-30
Beji, Lotfi II-30
Bellafkih, Mostafa I-246
Belmekki, Abdelhamid I-233
Benallou, Imane I-175
Bennani, Samir I-106
Bermeo, Pither Gabriel Tene II-3
Bouaicha, Mohammed II-248
Bouhsaien, Loubna I-175
Boukir, Khaoula II-215
Boukrouh, Ikhlass I-31

Bouskour, Sara II-120
Briouya, Asmae II-187
Briouya, Hasnae II-187

C

Chafik, Samir II-131
Chakir, Asmae II-107
Chakkor, Saad II-43
Choukri, Ali II-187
Chriss, Marouane I-259
Copeland, Jacob II-3
Coulibaly, Korota Arsène II-16
Curea, Octavian II-16

D

Daaif, Abdelaziz II-200
Dkiouak, Aziz II-43
Dornaika, Fadi I-3, I-121

E

Eddarhri, Maria I-153
EL Maazouzi, Qamar I-106
EL Moudden, Ismail I-46, I-74
El-Alami, Adnane II-238
Elallam, Oussama I-268
Elguemmat, Kamal II-200
Elkamouchi, Rahma II-200
Elkardaboussi, Youssef II-248
Elmoutaoukkil, Abdelwahed I-259
Enaanai, Adil I-46
En-Naimi, El Mokhtar I-204
Ennaji, Mohamed II-55

F

Franck, Dufrenois I-190

H

Haddi, Adil II-173
Haidoury, Mohamed II-55
Hain, Mustapha I-74, I-153

Hamim, Mohammed I-46, I-74
Hamlich, Mohamed II-16, II-70
Hammoudi, Karim I-3
Hidila, Zineb II-70
Hillali, Youness II-131
Houda, Taha II-30
Houmairi, Said II-248

I
Idiri, Soulaimane I-139

J
Jadli, Aissam I-46
Jami, Oussama I-268

K
Karaouni, Malak I-121
Khalfaoui, Hafida I-139
Khalfaoui, Samira I-31
Khamayseh, Faisal I-60
Khatib, Amine I-259
Kouihi, Manal II-93
Kourab, Zakaria II-55

L
Lagmich, Youssef II-43
Lemssaddak, Sofia II-145
Llaria, Alvaro II-16

M
Majdoul, Radouane II-238
Marzak, Abdelaziz I-153
Meddaoui, Anwar II-85
Mellal, Ilyasse II-215
Mellouli, Hala II-85
Mohamed, Tabaa I-190
Mouchtachi, Ahmed I-259
Moujoud, Loubna I-233
Moutachaouik, Hicham I-46, I-74

Moutchou, Mohamed II-93
Mouttalib, Houda II-159

N
Ndama, Oussama I-204
Nisrine, Bajja I-190

R
Retbi, Asmaâ I-106

S
Sabbar, Hanan I-17
Saidi, Rabiae II-16
Saim, Mohammed Marouane I-218
Senouci, Benaoumeur II-3
Silkan, Hassan I-17
Slika, Bouthaina I-3
Souabi, Sonia II-107

T
Tabaa, Mohamed II-70, II-107, II-145, II-159
Taleb, Ismail Ait II-55
Tayane, Souad II-55

V
Ventura, Sebastián I-161
Voskergian, Daniel I-60

W
Wakrime, Abderrahime Ait II-215

Y
Youb, Ibtissam I-161
Youssfi, Mohamed II-159

Z
Zaggaf, Mohammed Hicham II-120
Zaki, Abdelhamid II-85
Zaki, Mohamed I-268
Zegrari, Mourad II-131, II-145, II-248

Printed in the United States
by Baker & Taylor Publisher Services